COLLAPSE

COLLAPSE

When Buildings Fall Down

PHILLIP WEARNE

NEW YORK

Library of Congress Cataloging-in-Publication Data
Wearne, Phillip.
  Collapse : when buildings fall down / Phillip Wearne.
   p. cm.
  ISBN 1-57500-144-6
  1. Structural failures. 2. Building failures. I. Title.
  TA656. W415 2000
  690—dc21
                                      00-039230

Picture Credits
p. 29: Bob Campbell; p. 41: James Bashford, courtesy of the Gig Harbor Peninsula Historial Society; p. 49: West Virginian Archives; p. 52: John Bennett; p. 69: *Lancashire Evening Post;* p. 83: Darlow Smithson Productions; p. 87: Wayne Lischka; p. 92: *Kansas City Star;* p. 101: Dan Kook University; p. 106: Dan Kook University; p. 118: Darlow Smithson Productions; p. 121: Popperfoto/Reuters; p. 128: Popperfoto/Reuters; p. 151: Popperfoto; p. 165: Popperfoto/Reuters; p. 184: Sid Brown; p. 216: Popperfoto; p. 243: Norsk Oljemuseum.

TV Books, L.L.C.
1619 Broadway, Ninth Floor
New York, NY 10019
www.tvbooks.com

Interior design by Rachel Reiss
Illustrations by Donna Froude
Manufactured in the United States of America

CONTENTS

DEDICATION

THE DAY I FINISHED WRITING THIS BOOK, THE OFFICIAL DEATH TOLL for the August 17, 1999, earthquake in northwest Turkey reached 17,118. This book is dedicated to those victims and to all others who were in the wrong place at the wrong time during a construction disaster.

ACKNOWLEDGMENTS

I AM NOT AN ENGINEER, STILL LESS A STRUCTURAL ENGINEER, AND RE-ceived much guidance through the technicalities and terminology that were essential to an understanding of the subject of this book. Particular thanks to Sam Webb in Kent, England; Scott Steedman in Reading, England; Wayne Lischka in Kansas City; and Frank and Wendy Patton in Vancouver, Canada.

The material on which the book is based was collected for The Learning Channel television series "Collapse." I am thus deeply indebted to the whole production team, in particular Greg Lanning, the series producer, and his impressive team. Juliana Challenor, Caroline Hecks, and Noddy Sahota all briefed me on their research, then read the text and corrected my errors. Hannah Wilson kindly helped me supplement the production team's work by rooting around some libraries, institutes and the Internet. Debbie Arnsby and Melissa Stratford kept me plentifully supplied with reports and transcripts; Paul Gardner responded with his renowned efficiency to requests to view archive film.

At Channel 4 Books my thanks go to Charlie Carman and Verity Willcocks, who handled the whole project from start to finish. My editor, Christine King, dealt with the vagaries of my narrative style and structure with admirable patience and pro-fessionalism. And the illustrator, Donna Froude, has made plain what I often struggled to express in words.

INTRODUCTION

I T'S A PLEASANT, INNOCUOUS BUILDING IN THE AFFLUENT CALIFORNIA
suburb of Menlo Park. Landscaped with tidy gardens and
fountains, it looks sturdy, dependable, safe as...well, houses.
And so it should. For this is the home of Exponent or, as it was
more descriptively known at the time of the cases featured in
this book, Failure Analysis Associates (FaAA).

Here, surrounded by the hi-tech that is now so integral to
their profession, work some of the world's most experienced
"failure busters," as the U.S. press likes to label them. Transport accidents, building collapses, industrial disasters...FaAA
employs some of the best-known analysts in the world. Cases
can take up to ten years of work and, at any one time, FaAA has
a couple of thousand investigations going on.

Many of the case studies involve re-enacting an accident at
the company's 160-acre proving ground in Arizona; all of them
involve meticulous checks in a 350-million-accident database
that grows daily. It's a clinical, scientific system. "We just analyze the facts, the sequence of events, and present them to the
client," says Roger McCarthy, FaAA chairman and chief technical officer. "We don't make the judgments."

Founded more than thirty years ago, FaAA has become one
of the world's leading centers of what has developed into a
huge worldwide industry. Under its real name, forensic engineering, failure investigation in the construction industry

9

sounds glamorous. In reality, as anyone at Menlo Park will tell you, it is often a dirty and dangerous business.

It starts with a painstaking examination of the rubble of collapsed buildings, looking for clues. At times, sifting through the debris is not unlike an archaeological dig. "You excavate down through the layers and you can go back through time, back through the sequence of failure," says John Osteraas, a leading structural engineer with FaAA. It involves scrutinizing debris from every facet of the building: the columns, slabs, and beams that form its skeleton; the panels, bricks, and glass that form its skin; the foundations that are its feet.

In seeking answers to the question as to why a building has failed, forensic engineers will want to know its complete history. They have categorized seven stages at which potential failure could have been built into a structure—a building's equivalent of the seven ages of man. They range from the design stage when a building is conceived, to the construction when it is born, through its functional operation, when it is alive.

The clues unearthed on site, the design specifications of the building, the original calculations—all the relevant information will be collated. Computer models can be invaluable. Specialist expertise is often vital. Even then, with everything stirred into the pot, any solutions may only be theoretical. As with all detective work, definitive conclusions can be impossible. "Sometimes you can see immediately what went wrong, sometimes you could pore over the problem for a lifetime and still never be sure. You just never know," says one forensic engineer.

Structures, logically enough, are the preserve of structural engineers, who are engaged in what has been defined as "the science and art of designing and making, with economy and elegance, buildings, bridges, frameworks, and other structures so that they can safely resist the forces to which they may be subjected." When those forces prove irresistible, it is other structural engineers—forensic engineers—who try to determine why.

IN THE CONSTRUCTION INDUSTRY, FAILURE HAS TWO DEFINITIONS. ONE IS that generally understood by civil engineers: the unacceptable difference between "expected and observed performance." By this definition, failure in structures is very common. It encompasses everything, all forms of distress, deformation, or deterioration, from the leaking roof of the newly built garage to the cracked façade of the office building. In some major or minor way, the building simply did not meet its specifications.

The other definition is the failure analyzed in this book: catastrophic structural collapse. There are, thankfully, relatively few such failures—so few, in fact, that this book has had to take the broadest possible definition of "buildings" to encompass all structures. Bridges, an oil rig, and even an underground valve house are included here.

In answering the question of why a building failed, forensic engineers often come up with a combination of causes. And even if there is a single cause of a catastrophic collapse, it often has a knock-on effect, spawning the chain reaction that structural engineers term "progressive collapse." Thus a relatively small gas explosion caused the collapse of twenty-two stories of apartments at Ronan Point in London (see Chapter 7), and one minuscule flaw in the metal of one of the giant eyebars of a suspension bridge set off a chain reaction that brought down the Point Pleasant Bridge over the Ohio River (see Chapter 2).

Usually, the causes are human. Mistakes, misunderstandings, incompetence, ignorance, dishonesty: every facet of human failing is represented in construction failure—just as it is in any other branch of life. "Acts of God"—earthquakes, hurricanes, tornadoes—are common enough. The reality is that buildings are invariably designed to withstand them, whatever insurance companies may say. The materials used to construct buildings follow the laws of physics and nature exactly. It is human misunderstanding of how they work or failure to use them properly that invariably causes disaster.

The faulty weld that attached the underwater microphone to the main brace of one of the five legs of the *Alexander Kielland* oil rig (Chapter 11); the change of design of the fixings that held in place the skywalks at the Hyatt Regency Hotel in Kansas City (Chapter 4); the failure to protect the Thruway bridge over Schoharie Creek in New York State against the scouring effect of the river (Chapter 9)—these were all human errors.

Errors of knowledge—ignorance; errors of performance—carelessness and negligence; errors of intent—greed and corruption: all these and more are represented here in this book. The structural engineers who made them knew all the basic laws and rules of their profession. They understood that nature, whether wind, rain, sun, or snow, will always find the weakest link in a structure and work on it mercilessly. They understood that all structures, whatever their shape or height, must take on gravity and win—all day, every day.

These structural engineers were intimate with the secret world of loads—not just dead loads, live loads, or dynamic loads—but the invisible thermal loads caused by changes of temperature or suction loads caused by wind. They understood stress and strain: the tension exerted on a building by pulling forces or the compression experienced as a result of a pushing force. They understood how the structural materials they use would react when exposed to all these forces: their ductility—the ability to bend, stretch, and twist without breaking; their elasticity—the ability to return to their original shape once a load is removed; and the opposite, their plasticity—the inability to return to their original shape. Yet still, despite all this knowledge, in each case chronicled in this book something went drastically wrong.

But one thing is very clear: if success in structural engineering is avoiding failure, success stems most completely from learning from failure. "Structural engineers have only two methods of building in success: they can prove a structure is safe by means of calculation or they can build a replica and test it," one architect told me in the course of research for this book. "The method of choice open to most other designers is not open to structural engineers: they cannot build a prototype and test it to destruction. That's not an option."

As a result, whether arising from acts of man such as the bomb attack on the Alfred P. Murrah building in Oklahoma City (Chapter 6), or acts of God such as the earthquakes of San Francisco and Kobe (Chapter 8), failed buildings are invaluable field laboratories for structural engineers. Failure is in fact the only basis for successful design. Disaster is most likely when new designs are based purely on successful precedents and the basic lessons of past failures are ignored. There is no better example of such complacency and over-confidence than the construction and collapse of the Tacoma Narrows Bridge sixty years ago (Chapter 2).

If we learned and applied the lessons of each failure, the structures in our world would be getting safer all the time. But that is a big assumption. And other forces are at work. New materials, competitive pressures, economic constraints are all exerting pressures on architects and structural engineers. Everywhere these professionals are pushing the envelope of the possible. New techniques allow them to go higher, faster, and wider with new materials that must by definition be less tried and tested. New economic pressures encourage them to "refine" designs to the point where long-accepted margins of safety are breached.

Taller skyscrapers, wider atriums, ever lighter spaces—today's buildings are no longer just bricks and mortar but complex machines, bending, stretching, and pulling materials to unprecedented limits. Designed by computer, buildings are increasingly run by computer. But although we are nearing the time when your house can ring you on your mobile phone to ask you when the central heating should be switched on for your return, we are still some way from the time when structural components of a building will be able to trigger alarms warning us they are about to fail.

And therein lies the problem for structural engineers. As President Herbert Hoover, a former engineer, remarked, engineering is the most unforgiving of professions. "He [the engineer] cannot bury his mistakes in the grave like the doctor... argue them into thin air or blame the judge like the lawyer. He cannot, like the politicians, screen his shortcomings by blaming his opponents," Hoover claimed. "The great liability of the engineer is that his works are out in the open where all can see them. The engineer simply cannot deny he did it. If his works do not work he is damned."

1

FAILURE
BY DESIGN

THE ROOF FELL IN . . .

Hartford Coliseum and Kemper Arena

NO CONSTRUCTION FEATURE CAUSES MORE PROBLEMS THAN ROOFS. From the odd missing slate or tile that lets in rainwater on a domestic roof to the complete collapse of space frames covering vast indoor arenas, roofs—whether leaking, rotting, moving, collapsing, or being torn off by hurricanes—are perhaps the most vulnerable part of any structure. One architectural periodical calculated that in the 1980s roofing, a mere 6 percent of total construction costs, accounted for nearly 40 percent of liability claims made against architects. "It takes both knowledge and perseverance to get a roof right," one developer told *Engineering News-Record*. "Anyone who hasn't had a major roofing problem is either lying or lucky."

Perhaps because roofs *are* such a problem there has been more innovation in roofing in terms of materials and design over the past thirty years than in perhaps any other feature of the construction business. The explosion in the use of these innovations in roofing has created its own problems—the most obvious one being the absence of useful statistical data on the performance history of both designs and materials. In the 1970s and 1980s chemical companies developed a wide variety of new membrane roofing materials including modified bitumen products, a rubber-based substance known as EPDM, and polyvinyl chloride or PVC, a plastic. New materials in turn had an impact on new designs, making innovation both possible and common.

Unique, high-profile structures have also been using new designs even if at times it's with older materials. The most spectacular of these have been modern dome-shaped roofs spanning distances of up to seven hundred feet with reinforced concrete and curved lattices of steel beams. The Superdome in New Orleans is the best-known example. Another new design has been tensile roofs. Based essentially on the principle of a tent, they use the pull of tension from columns, walls or radial steel cables to suspend circular or elliptical rings of compressed concrete. Tensegrity domes, a combination of the two designs, were developed in the 1960s, and in 1990 the world's first elliptical (as opposed to circular) tensegrity dome was designed by Matthys Levy to cover the Olympic Stadium in Atlanta.

But the most commonly used modern roofing innovation to enclose large spaces is the space frame. The basic component of a space frame is a truss, a steel triangle composed of parallel bars or chords top and bottom connected by struts running at angles. The two components are connected to form one of the stablest shapes a structure can have. Space frame roofs, now universal in sports stadiums, shopping enclosures, and warehouses, are essentially three-dimensional combinations of parallel trusses (see Figure 1, top). Light, economic, and elegant, space frames can move monolithically in two directions with the bars or chords absorbing the tension stresses and the struts running between them the compression stresses.

Yet even the most modern roof design is vulnerable, and at no time were the consequences of gravity, loads of water, or snow on roofs more graphically illustrated than in two winters in the United States. More than 130 major collapses were recorded in record snowfalls across Wisconsin, Michigan, and New England in 1977–78. The following winter more than 200 major roofs collapsed in Chicago alone when the city suffered forty-seven inches of snow in one month. Snow loads, as the city's engineer described them, reached fifty pounds per square foot and more. College roofs, part of a shopping mall, a truck-loading dock, a school gymnasium roof, and warehouse roofs all came tumbling down. City and state building codes carried specifications that were simply insufficient to withstand the weather.

Nowhere was there a collapse more dramatic than at the Civic Center Coliseum in Hartford, Connecticut. It happened just after 4 A.M. on January 18, 1978. There were no casualties and few witnesses—but along with the hotel night clerks and car-parking attendants, who said they heard a rumble then a boom, was Horace Becker. He was a guest at Hartford's Sheraton Hotel, staying in a room facing the Coliseum. Awoken by a loud "cracking noise," Becker recounts how he went to the window in time to see the corner of the Coliseum roof rise, then the center sink, with what he describes as "a whooshing sound."

When the windows of his room began to shake, Becker thought an aircraft had crashed into the building. He dropped to the floor and did not dare look out of the window until the vibrating and crashing had stopped. When he did peer out over the window ledge he saw that all four corners of the roof were pointing up into the night sky. The two-and-a-half-acre roof had collapsed in the center, pancaking eighty-three feet onto the twelve thousand seats below. Some fourteen hundred tons of steel, insulation material, and roofing panels had fallen into a space where five thousand spectators had been watching a basketball game just six hours earlier.

The prime suspect was the weather. Heavy snow had fallen that night for the second time in a week. Tests by the Army's cold regions research and engineering laboratory showed that snow on buildings across the street from the Coliseum weighed twenty-three pounds per square foot, although there was reason to believe the snow load on the roof itself was

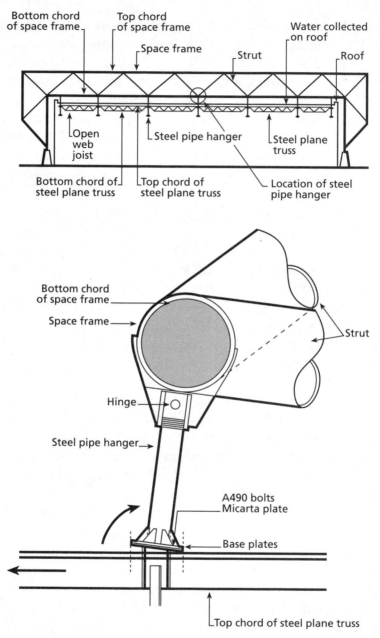

FIGURE 1

Top: A cross-section of a space frame roof. *Bottom:* Detail of steel pipe hanger at the Kemper Arena, Kansas City.

lighter, just ten to sixteen pounds per square foot. Neither figure seemed to matter. The design specifications from the Hartford offices of Fraoli, Blum & Yesselman, the engineers who had constructed the Coliseum, showed that the roof was designed to withstand live loads of thirty pounds per square foot for snow and twenty pounds per square foot for rain.

There was one obvious question: were the design specifications adhered to? City officials appointed a three-member city council panel to handle the investigation. The panel duly appointed Lev Zetlin Associates (LZA) of New York City to lead the investigation. On the face of it, the Coliseum roof was a standard space frame—there were dozens like it throughout the United States. In reality, the space frame roof of the Hartford Coliseum was anything but standard—it had gone through a number of adaptations.

In 1971, the design engineers for the Hartford Coliseum proposed an adapted space frame roof structure that they claimed would save half a million dollars in construction costs but would require a complex and expensive computer analysis to check. The city council gave the go-ahead and the computer-generated model was deemed to meet all the design-load criteria. One of the key differences in the modified design was that the roofing panels, which normally rested directly on top of the horizontal chords of the space frame, were supported on short vertical posts above the top nodes of the space frame (see Figure 1, top). The idea was that the height of the posts could be varied to provide a gradient for better drainage. The posts would also protect the top chords of the frame from the bending stresses of roof loads such as rain or snow.

The investigators quickly concentrated on the configuration of the actual structure as compared to the mathematical computer model registered by the designers. They found what they considered to be deficiencies, errors, and false assumptions. The most important of these was bracing. The space frame consisted of a series of thirty-foot-square bays. The original design calculations provided for an unsupported length of fifteen feet for each member or segment of top chord, in other words a brace at the mid-point. Such bracing was crucial in a space frame—the load-bearing capacity of a relatively slender compression member in such a structure will always depend on how and where it is braced.

Investigators quickly established that in the Hartford Coliseum space frame roof, the top chord compression members were not braced at all. The mid-point bracing had just been omitted in construction, leaving the exterior top chords unbraced for their entire thirty-foot length. Compounding the problem, the interior top chords were only partially and inadequately braced at their mid-points. Charles H. Thornton, heading LZA's investigation, was at a loss. "Where and how the omission occurred is hard to say," he complained to journalists.

The overload resulting from this omission was staggering: 852 percent in the exterior top chord compression members on the east–west faces; 213 percent on the corresponding exterior chords on the south and north faces; and 72 percent on the interior top chord compression members in the east–west direction. Buckling, bending, fracturing, and eventual failure of the top chord members was inevitable, investigators decided. The true load-bearing capacity of these steel bar members was only 9 percent of that assumed in design. Future failure had been built into the structure.

Errors in the calculation of dead-weight loads were compounded by other deficiencies built into the structure during the construction process. Some of the space plates at the four-angle cruciform members were placed too far apart, causing buckling early on. Some of the steel used did not test to its specifications and some of the diagonal members were actually misplaced in the space frame. Most obvious of all was a change of roofing materials by the construction manager during building. That decision alone increased the design space-frame weight from eighteen pounds per square foot to an actual weight of twenty-three pounds per square foot.

Perhaps most extraordinary of all was the apparent blind reliance on the theoretical computer-model outcome in the face of overwhelming evidence that in practice something was very wrong. Obvious problems were ignored. Correspondence, queries, complaints from everyone from official inspectors to concerned citizens were simply ignored or fobbed off. People were constantly assured everything was all right. Typical was the photograph produced by LZA investigator Thornton indicating bowing in two top chord members during the fabrication of the truss. "This amount of bowing should have been a red flag indicating there was a problem," he insists.

It was only one of many clues. Just one more of the unique features of the Hartford Coliseum space frame roof was that it was assembled on the ground. During the four months the structure was being bolted together, engineers were notified by the inspection agency of a suspicious and excessive deflection of some nodes or joins. Nothing was done; after all, the computer calculations had said it was safe. In January 1973, the space frame was lifted into position by means of hydraulic jacks without the actual roof deck. At this point, the space frame was actually measured as having a central deflection twice that predicted by the computer analysis. When notified, the engineers changed tack. Such discrepancies were to be expected in view of the simplifying assumptions made in the theoretical calculations, they argued. The logic of their own argument demanded a physical check. But still nothing was done.

Within days of this, the subcontractor fitting the fascia panels covering the façades of the space frame on its four sides complained. When he tried to mate the pre-punched bolt holes in the two pieces of steel of the space frame with those in the façade panels the holes were way out of line. It was another indication of severe deflection—what structural engineers term "random boundary deflections." The subcontractor actually wrote to the Coliseum's general contractor, Gilbane Building Company, outlining his worries. In reply, Gilbane told the subcontractor to deal with the problem or be responsible for the delays. Gilbane told LZA's investigators that they themselves had been told that whatever deflection occurred was to be expected.

In mid-1974, with the Coliseum now completed, a member of the public expressed concern about a "large dip" in the roof. Both the engineer and the contractor assured city officials that there was no cause for concern. Finally, in January 1975, just days before the Coliseum's official opening, a Hartford councillor revealed that a construction worker had told her that the deflection of the roof was almost twice that originally calculated. This time the city did not even refer it to the project's engineers, but took their own measurements. They confirmed the deflection. Yet still nothing was done.

One problem at the Hartford Coliseum construction project had been "the absence of a full-time registered structural engineer experienced in the design of long-span special structures," according to LZA's final report. Neither the project architect nor the structural engineer for the Coliseum was retained to keep an eye on things during construction. It was not for want of trying. The citizens' panel appointed by the city to look into the disaster discovered minutes of two project meetings during which the architect tried to get a qualified structural engineer appointed. He wanted continuous on-site inspections during the space frame assembly and construction.

The idea was rejected by the construction manager as a waste of money. He took on all the inspection responsibilities himself. Thus, at a stroke, the construction manager eliminated the system of checks and balances that had proved so vital on construction projects in the past. On site he became responsible for design, workmanship, and materials—effectively inspecting his own work. As the citizens' panel noted rather euphemistically: "inspection of his own work by the construction manager is an awkward arrangement."

The central lesson for the Hartford Coliseum failure was simple: the computer is only an analytical tool and can only be as accurate as the figures and assumptions keyboarded into the model projection. Assumption was of course no substitute for the reality of actual measurements in the field. It was an important early lesson at a time when the use of computers was about to take off in the design and structural engineering fields. It was ironic, then, that it was computer modeling that ultimately gave investigators their failure sequence. Even more ironic, perhaps, that the same modeling also demonstrated the allowances that needed to be made when designing by computer.

Armed with the correct buckling lengths and stiffness of all the chords in the space frame roof, the computer model was programmed to discover at what load the first chord would buckle. It was conservatively estimated at 13 percent below the estimated roof and snow weight on the day of the disaster. As expected, the model showed that when one chord buckled the load was transferred to adjacent chords that could not carry the load and then themselves buckled. In other words this was a "progressive collapse" based on the structure's lack of redundancy, or put another way, the building's simple inability to successfully transfer the load resulting from one failure elsewhere.

The computer model demonstrated in graphic detail how as each bar of the space frame fails, more load is transferred to fewer bars until the whole roof collapses in the sort of inward folding pattern observed at the scene of the disaster. The "live collapse load"—the weight of snow under which the whole roof collapsed—was put by the computer at 20 percent above that actually believed to be on the roof at the time of the disaster. This difference between the real weight and the theoretical weight seemed to demonstrate the problem of the use of computers in structural design: no real structure can be as precise or perfect as the computer model.

THE CIVIC CENTER COLISEUM IN HARTFORD WAS REBUILT. BUT THE PROBLEMS with space frame roofs elsewhere were far from over. Indeed, within eighteen months there was another major roof collapse halfway across the country in very similar circumstances. Weight on the roof was again the initial cause; progressive collapse again the result. Cause and effect were linked by design deficiencies, errors, and shortcomings that collectively added up to the same thing—failure. The question was:

could the forensic engineers who now stepped into the debris read the trail of clues and signposts left behind accurately enough to come to as comprehensive a conclusion as they had done at Hartford?

Kansas City has some of the most solid-looking buildings you are likely to see anywhere in the world—substantial, rectangular structures of red and brown sandstone that look as though they would survive Armageddon. Perhaps it is something to do with the city's transitory origins. Historically Kansas City—the twin city at the confluence of the Missouri and Kansas Rivers, partly in one state, partly in the other—was a rest-stop, then a settlement on the main trails west to California and Oregon. Neither east nor west, a sort of halfway house on the road someplace else, both parts of Kansas City have worked hard at putting themselves on the map. Galleries, a museum of the arts, a symphony orchestra, and a respected university—all of which needed impressive home bases—were part of the endeavor. So were the Kansas City Royals, the baseball team, and the Kansas City Kings, the local basketball franchise.

In 1973, the Kansas City Kings had a new state-of-the-architect's-art stadium built on the site of the city's old horse and cattle fair. Named after R. Crosby Kemper Jr., one of the city's founding fathers, it seated 17,600 and was completely covered by a huge roof deck. Hung from a space frame erected outside the arena, the Kemper Arena managed to achieve the holy grail of stadium design. With no internal supports or pillars, every seat in the stadium had an unobstructed view in the 324-foot sweep of internal open space. But it was not only supremely functional—the Kemper Arena looked good, too.

"Sleek and futuristic" was how *Time* magazine described the three huge exterior trusses or interlocking network of pipes that "marched up, across and over the cool white structure, holding up the roof and giving the building a light, lacy effect." The Kemper Arena was so innovative that in 1976 its designer, Helmut Jahn, and his firm, C. F. Murphy Associates, were lauded with the Honor Award of the American Institute of Architects (AIA). Indeed, on June 4, 1979 the AIA opened its annual convention in Kansas City. En route to their venue about a mile from the Arena, dozens of AIA members stopped to study the city's architectural showcase. They all agreed it was impressive: four acres of tin roof hanging from the triangular steel space frames, lightness and strength all in one.

As they left to attend the opening of their conference, AIA members could not have known that within hours both these features would be tested to the limit. Both would be found horribly wanting. At six o'clock that evening the skies darkened; by six-thirty a torrential rainstorm had begun. A little more than thirty minutes later, hearing a loud "thundering noise," Art LaMaster, the stadium's supervisor, went down to the stadium floor with a security guard. Two massive streams of water were pouring from the roof onto the floor, falling on either side of the stadium's eighteen-ton scoreboard, itself suspended from the roof.

The water continued pouring down for three or four minutes. Then something seemed to give way. "There was a roar like the pounding of a sledgehammer on concrete then the scoreboard broke loose in the center," LaMaster recalls. "I didn't watch any more. I was too busy getting out of there. I just ran for my life." Art and the security guard had about five seconds before the central portion of the roof—a huge piece about 200 feet by 215 feet—crashed onto the stadium floor and seats in a mass of twisted steel, broken glass, and insulation material. Just two nights before, more than thirteen thousand people had been seated there attending a truck drivers' convention. Three days later, seventeen thousand were due in for a Rod Stewart concert.

The AIA had opened its annual convention across town earlier that afternoon with the presentation of another award to Helmut Jahn—this one for the design of a gymnasium in South Bend, Indiana. He heard the news of the Kemper Arena roof collapse at the AIA's opening banquet. "It's just terrible.... The building was designed to withstand certain winds and certain conditions but there are such things as acts of nature," he proffered in explanation. City officials were skeptical. "This is tor-

nado country," said one. "We've seen worse than that. The
storm caused no other major damage."

Indeed, weather records showed that in the six years it had
been standing the Kemper Arena roof had withstood much
worse wind and rain. Frederick P. Ostby, deputy director of the
National Severe Storms Forecast Center in Kansas City, confirmed that storms with winds of the same severity occurred
"about once a year." This, it was clear, was not one of the
"hundred-year floods" or "hundred-year rains" that structural
engineers were so often inclined to blame for their failures.

Don Hurlbert, the city's engineer, recalls heavy rain but
nothing exceptional. What did stick in his mind was the
wind—seventy to eighty miles per hour at times—with rapid
changes in atmospheric pressure. His main concern that night
was tall buildings: he was getting reports of people coming
down off high control towers because they were swaying so
violently. About 8 P.M. he got a call from the city architect responsible for maintenance. "He said he was down at the Kemper and there was a hole in the roof. I said, well, it's too late to
do much, let's put a tarp over it and we'll look at it in the morning. He said: 'Mister, there's not enough tarp in Kansas City to
cover this hole.'"

Don Hurlbert got down to what had been the Crosby Kemper
Memorial Arena first thing in the morning. Bizarrely, the space
frames were still in place; in fact they were virtually undamaged. From the debris on the arena floor it seemed that the failure had occurred in the connections between the space frames
and the secondary roof structure spanning between them. This
secondary roof structure, which was essentially steel plane
trusses, had been suspended from three space trusses by steel
pipe hangers at forty-two different panel points (see—Figure 1,
bottom, p. 20). In falling, the central section of the roof had
acted like a giant piston, exerting so much air pressure that
some of the walls of the arena had been blown out.

There was no shortage of architectural expertise or opinion
in Kansas City. Half the AIA members attending their conference were soon picking over the carcass of the building. "It had

The space frame roof of the Kemper Arena in Kansas City that folded in the middle after the weight of "ponding" water proved too heavy a "dynamic load."

to be something like a vibration, an oscillation," insisted Helmut Jahn at the site, convinced the wind and rainstorm were to blame. Don Hurlbert favored fluctuations in air pressure. Perhaps a blown-out window might have caused more pressure to build up under the roof than above it, literally blowing the roof upward before gravity had brought it literally back down to earth.

But most experts opted for something to do with the rain and wind. Water, after all, was the most common cause of the most common failure in structural engineering—roof collapse. The urgency to get to the bottom, or perhaps more accurately the top, of this problem was compounded by the need to restore public confidence in the wake of the Hartford Coliseum roof collapse. "When you put two things like that back to back it's got to have a tremendous impact.... Public confidence has been shattered," observed Thomas H. Teasdale, a newly elected AIA vice-president. "It doesn't help the public's respect for architects," conceded John Sheehy, another AIA member in Kansas City. "But the real question here is where does the responsibility lie?"

That was a big problem. To an even greater extent than the Hartford Coliseum, the Kemper Arena had been built to a non-traditional design—it was new, experimental, unique. It was what was known in the trade as a build-contract: the designer worked under a direct contract with the builder, effectively as a subcontractor, rather than under the traditional owner-architect agreement. Analysis of the causes would have a huge impact on the legal quagmire that now beckoned, with litigation likely between builders, subcontractors, material manufacturers, owners, designers, and the municipal authorities. As a result, all parties involved wanted their own investigators. And not only did the causes have to be established, the Arena had to be rebuilt—and fast. Kansas City had a contract with a team playing basketball here and they were threatening to move.

Don Hurlbert and the city authorities appointed James L. Stratta, a civil engineer and a widely respected failure analyst, to investigate the causes. J. E. Dunn Construction Company of Kansas City, the contractor on the job, hired civil engineer Bob Campbell to look at the collapse from their point of view. A subcontractor hired another firm, Weidlinger Associates of New York; and the steel manufacturer KC Structural Steel, facing allegations that their joints holding the roof in place had failed, hired Failure Analysis Associates (FaAA). Division and disagreement would mean that it would be 1983 before the designers, the builders, and the city would agree to share the six-million-dollar rebuild cost, a compromise which suggested that no single party could be held responsible.

Stratta's investigation began with the self-supporting roof frame. Visual inspection showed no defects. However, the remains of the bolted connections that formed the link between the suspended roof and space frame did. "The bolts broke at the root of the thread," Stratta observed just days into his inquiry. "It appears to be a direct tension failure. They pulled apart." So was it the bolts themselves? Stratta was cautious but even at this early stage he seemed to suspect otherwise. "The bolts may have been a secondary failure...they may have failed because of something else. We don't want to get into a DC-10 problem," he warned a reporter from *Engineering News-Record,* referring to a recent airline crash, "where investigators said the sheared bolt created the problem and then a week later said the problem created the sheared bolt."

A collection of the one-and-three-eighths-inch A490 friction bolts that passed through the steel plates to hold the roof in place were sent off for stress testing. They met all their design specifications with ease. That indicated another cause, the most obvious candidate being a tremendous downward force on the roof. But that in itself could have had numerous causes: differential wind pressure inside and outside the structure, ponding (the collection of rainwater), vibrations, or structural defects. It might even be a combination of some or all of these.

For Roger McCarthy of FaAA the investigation had to start with the weather conditions. "The nice thing about a structural collapse in a rain- and windstorm is that things collapse because the load was too great or the structure didn't have enough strength—which is of course the same thing. So there's no question the reason this roof came down is that the loads on it exceeded the strength of the material to resist it. That wasn't a mystery." But what the load was and why it had been too much did remain a mystery.

There had been speculation that some or all of the eight drains on the roof had been blocked, allowing the water to pond or accumulate in heavy pools and stress the roof beyond its weight-bearing capacity. In fact, none of the drains were blocked. Yet investigation and detailed calculation showed that ponding would have occurred on the roof on June 4. In fact, ponding was designed to take place.

The weathermen said that some three and a half inches of rain fell on the Kemper Arena roof in the first hour of the storm which started just before 6:30 P.M. That amounted to twelve hundred tons of water falling on the 131,440-square-foot roof. The only outlets for the water were the eight drains which, rather than being on the edges of the roof, were actually toward the middle—placed between the middle and outer space trusses, four to a side. To enable the water to run to these drains, the roof had an inverted pitch, a six-inch incline from the outer edges to the drains.

These drains were known as "flow control design"—in other words, they did not let too much water out at once. Using roofs as temporary reservoirs by limiting the flow of rainwater in the event of sudden downpours was not an unusual feature in rapidly expanding towns with overloaded sewer systems. On the Kemper Arena roof this was achieved by limiting the diameter of the pipes: four-inch pipes on the roof became five-inch pipes and finally eight-inch pipes as they led into the actual storm sewers. This deliberately prevented each drain from discharging more than one-tenth of a cubic foot of water per second.

The only other outlets for water were scuppers through which water could fall directly to the ground from the perimeter of the roof in an emergency. But the scuppers were set two inches above the roof level. Thus before water could force its way through them it would have to be two inches deep at the edge and six inches deep at the lowest point of the roof gradient. Yet although that was a lot of ponding, the roof had been designed to take such weight. Calculations by investigators showed that the weight and depth of water that would have accumulated in the June 4 storm was well within the design load-bearing capacity of the roof, as well as that of the plates and the bolts that held it in place.

Attention turned to the steel of the roof, what Don Hurlbert jokingly called the tin can, and the steel of the base plates that connected the roof to the space frame. The base plates had remained in position over the collapsed portion of the roof but had been twisted and bent in all sorts of different directions. The steel of the roof itself showed stresses, strains, and fractures, all of which held vital clues for the investigators.

Roger McCarthy explains: "Think of the structural steel as being like photographic film. Thin steel, like that on the roof, would be like ASA 400 film—very fast. Thicker steel, like that of the base plates, would be much slower—say, ASA 25 film. Both record a photographic image—one takes a lot of energy to imprint an image on it, the other very little. Once that image is recorded it doesn't come out, it's embedded. You just have to know how to develop that image, to read and interpret the deformation bending patterns to figure out what happened in the collapse. Ultimately, there is only one failure sequence that will account for all the damage you can read in the structural steel."

Using what was left of the roof and the pieces of fallen wreckage, Roger McCarthy set about trying to develop the film print of the collapse. The first thing he noticed was the base plates were now left hanging free in space. They were all sorts of different shapes. "Some had actually had their ears pulled off where the bolt holes went through. With others, one lip of the plate had been bent down and the other end bent up," he recalls. "So it was clear that the base plates themselves had been subject to different directional loads, different levels of stress, and had responded differently."

The key was the flexibility of the building. The Kemper Arena was a triumph of modern architecture—new, light materials used more efficiently with the lowest possible margin between strength and load. Yet its strength was also its weakness. "It was open, more airy, much more light—inevitably there was greater flexibility in this structure than a traditional post and pillar building would have had," says McCarthy. And on the roof the flexibility was accentuated by the looseness of the bolts. They were never tightened sufficiently to bring all the structural plates in the connection—top and bottom—into full contact with each other. Testing at the University of Missouri in Columbia showed that a torque load of only 400 to 750 pound feet was necessary to loosen the bolts. It should have been more like 2800 pound feet.

The Kemper Arena roof had been able to flex but perhaps it had been able to flex too much. "Sometimes we get too close to the margin between strength and load and we forget that there are really two things you have to worry about in a structure. One is the strength and the other is the stiffness," explains Roger McCarthy. "Sometimes the stiffness will govern the design. Even though you don't need the extra material for strength you may need the extra material for rigidity." The flexibility of both the loose bolts and light roof suggested that the Kemper Arena failure did have one very important thing in common with that at the Hartford Coliseum. The failure had begun the day the roof went up.

For not only had the water been able to pond on the roof of the Kemper Arena, it had been able to move around because the roof was so flexible. In wind and rain, the roof was effectively a live structure with a dynamic load. "The rain falls, the structure flexes a little bit and we get a pond. The pond on the roof creates more weight, structure flexes a little more. Pond gets deeper," explains Roger McCarthy. But then there was the complicating factor of the wind—gusts of up to ninety miles per hour. "You have a wind that can drive the water around and move the pond to the softest part of the structure. It's Mother Nature at her best—or worst—testing our structure with a dynamic water mass. And of course where the structure deflects most, the pond stops moving. That's the lowest point and the wind cannot blow it out."

In a vicious circle of increasing loads causing increased deflection, the roof became highly unstable. And all this had happened many times. In the six years the building had been standing, wind and water loading had caused a repeated rocking movement, subjecting the bolts to a dynamic loading they were never intended to carry. This motion fatigued, loosened, and weakened these high-strength bolts which, because they were relatively brittle, became even more vulnerable when subjected to dynamic loads.

Indeed, steel-design specifications expressly warned against the use of A490 bolts with dynamic, variable loads, recommending their use only when facing dead, static loads. "They [the bolts] really had deteriorated," James Stratta reported. "They apparently failed at between one-fourth and one-fifth of what they should have been able to carry. They were ready to go." For Stratta, the use of these bolts and their behavior in a moving environment was one of the most important mistakes made at the Kemper Arena.

"Structural engineers that I have talked to have been amazed to hear about the fantastic drop in tensile strength in these bolts after being jerked around....I don't think enough emphasis is placed on this." Many designers, Stratta complained, did not consider wind load a dynamic load and thus might mistakenly use the bolts in a dangerous application. "Anybody who has used these bolts now in pure tension situations yet with a possibility of dynamic loading introduced by wind should seriously consider replacing some and testing the rest for possible fatigue," he told *Engineering News-Record*.

It is the one area of Kansas City's official investigation with which Roger McCarthy disagrees. He does not deny that a good number of the bolts were broken but is convinced that they were not weakened before the disaster. "The bolts developed their full strength. They bent almost every plate they were bolted into. They didn't give up without a fight," he claims. "I don't care how strong a bolt it is, if you're willing to prise it with enough load you can break it."

Pointing to a sample of the type of plate and bolt that failed at Kemper Arena, Roger McCarthy demonstrates what he means. The more ponding, the more flex; the more flex, the more ponding, he motions with his hand. "Eventually those base plates with the bolts through them act like crowbars and a perfectly good bolt on an almost inch-thick crowbar got prised until it popped."

THE KEMPER ARENA ROOF WAS REBUILT WITHIN TEN MONTHS OF THE COL-lapse. James Stratta decided that the primary structural frame system was sound and that the roof could be rebuilt by simply redesigning and replacing. The hangers were modified and welded to the trusses, which in themselves were strengthened. The joists were deepened. But there were three other crucial changes.

Firstly, the four bolts, two steel plates, and the Micarta plate—a plate made of a resin containing plastic material—that made up the connections in the original design were replaced by a much more ductile single steel bar for each hanger, welded to the truss diagonals. Secondly, the roof was raised thirty inches in the center, reversing the inward slope of the original design, and fourteen drains were positioned at the perimeter of the roof. Thirdly, an electronic monitoring device warning of motion problems in the roof was added. It is programmed to be activated when winds reach thirty-five miles per hour.

To date, all the modifications have worked. For twenty years sports fans, music lovers, and political groupies have now been raising the roof rather than waiting for it to fall in on them at the Kemper Arena.

SUSPENSION BRIDGES

Galloping Gertie and the Point Pleasant Bridge

F ROM THE DAY IT WAS OPENED TO TRAFFIC, THE TACOMA NARROWS Bridge always undulated in the wind. Motorists had even complained of feeling seasick—crossing the center span could be like taking a ride on a giant roller coaster with the cars ahead liable to disappear in the trough of the tarmac waves coming at you. To the local people the bridge's nickname, Galloping Gertie, was thoroughly deserved. Then, after just eighteen weeks of life, Gertie twisted and jived so violently that it snapped.

On the face of it, Gertie was a conventional lightweight suspension bridge. In 1938, when construction work on the bridge began, the technology and science were hardly new. Indeed, the principle of hanging a roadway or path from towers or stakes by means of two main cables went back hundreds of years, back into pre-Columbian history in the Americas, when vegetable-fiber ropes and vines were the main materials. The only differences in the twentieth century were the scale and the materials used. The Tacoma Narrows Bridge, 11,440 tons of steel and more concrete than the constructor could total, was more than a mile long and hung from two 420-foot-high towers. The 2800-foot center span made it the third-longest suspension bridge in the world. Only the Golden Gate Bridge in San Francisco and New York's George Washington Bridge were longer.

But if the science of suspension was hardly new, neither were the failures. Lightweight suspension bridges had been particularly vulnerable historically. In 1826, just one year after its completion, Thomas Telford's Menai Straits Bridge linking the island of Anglesey with mainland Wales was torn apart in hurricane-force winds. Violent vertical undulations—some with wave heights of as much as sixteen feet—were its undoing. In 1854, a 1010-foot-span suspension bridge that crossed the Ohio River at Wheeling, West Virginia, broke up in the same way. One witness said it twisted and writhed and at one time the roadway "rose to nearly the height of the towers." Then, thirty-five years later in 1889, the 1260-foot Niagara–Clifton suspension bridge collapsed in winds clocked at up to seventy-five miles per hour. Witnesses said it rocked like a boat in a heavy sea.

Structural engineers thought they not only knew all about such movement but also understood the cause. John Roebling, designer of the Brooklyn Bridge, which was completed in 1883, had always based his work on the principle that the stiffness of the deck was absolutely critical to wind resistance. Stiffness was a function of two things. One: the effectiveness of the stiffening girders that prevent excessive sag when people or vehicles cross; two: the ratio between the length of span and the width of the roadway, what structural engineers term the span/width ratio. Throughout the twentieth century suspension bridges had been getting longer and sleeker, making them more flexible and potentially more vulnerable.

The Tacoma Narrows Bridge, connecting Seattle and Tacoma with the Puget Sound Naval Yard at Bremerton in Washington State, was no exception. In fact, it was the most slender suspension bridge ever built. The deck was only thirty-nine feet—it took two lanes of traffic compared with the six that crossed the Golden Gate Bridge and the twelve on the George Washington Bridge. In terms of span/width ratio, Tacoma Narrows had less than one-third of the stiffness of its two longer cousins, and the bridge's dead load was just one-tenth as much as that of any other suspension bridge in existence. The depth of the stiffening girders was just eight feet and the bridge was designed without any stiffening trusses for aesthetic reasons.

Yet for all this, the bridge's pedigree seemed impeccable. It had been designed by Leon Moisseiff, one of the world's leading figures in suspension bridge construction, a consulting engineer on the Golden Gate Bridge and many others. Moisseiff's design concepts and calculations were internationally respected. Some movement, he counselled, was to be expected. The Golden Gate Bridge itself had moved two feet up and down and six feet from side to side in recent windstorms. No one had suggested its structural integrity was threatened.

But even if remedial measures were needed at the Tacoma Narrows Bridge, Moisseiff had an unparalleled track record in making things right. It was his remedial measures on the Bronx-Whitestone Bridge in New York City—whose structure was similar—that had made the design flexibility of the Tacoma Narrows Bridge acceptable. Vibrations on the Whitestone Bridge had been reduced by the addition of an untuned dynamic damper—in essence a piston in a cylinder that absorbed much of the shock.

But the dynamic damper was just one of a series of remedial measures that had no obvious effect on reducing the undulations at Tacoma Narrows. Steel tie-down cables one-and-a-half inches thick—cables that effectively tied the roadway in each span to concrete anchors below the bridge—snapped in a moderate wind in the early autumn of 1940. In October, worried about the area's notorious winter gales, engineers put the bridge under permanent observation by means of a series of instruments measuring movement, stress, and the direction and speed of the wind. They also ordered experiments on a fifty-four-foot, one-to-one-hundred ratio scale model of the bridge under the supervision of Professor Burt Farquharson, associate professor of civil engineering at the University of Washington.

The engineers had three particular worries. Firstly, the scale of the oscillations and what set them off: certain winds seemed to set the bridge into a frenzy of movement completely dispro-portionate to the actual wind speed and the length of time they lasted. Secondly, the diverse nature of the movements: the in-struments on the bridge established that it could vibrate like a string in at least three different modes, the most alarming of which was a torsional or twisting motion; this, with the verti-cal undulations, added up to a sort of corkscrew twist in three dimensions. Thirdly, the duration of the movement: on other bridges that moved in this way, the oscillations seemed to die out or be dampened quite quickly. But not here at Tacoma Nar-rows—just another way in which Galloping Gertie seemed to be unique.

Some research on the model of the bridge, in particular the streamlining or shielding of the girders, was showing promise when a fierce wind hit the Narrows at dawn on November 7, 1940. As the wind rose—the anemometer on the mid-span reg-istered thirty-eight miles per hour, then forty-two miles per hour, then forty-five miles per hour—the bridge rose with it. Professor Farquharson was at the site early that day. Kenneth Arkin, the chairman of the Washington State Toll Bridge Au-thority, had already arrived, concerned about the wind. To-gether they closed the bridge to traffic shortly before 10 A.M. but not before Len Coatsworth, the local newspaper editor, had driven onto it.

"Just as I drove past the towers, the bridge began to sway vi-olently from side to side," Coatsworth later recalled. "Before I realized it, the tilt became so violent that I lost control of the car. ...I jammed on the brakes and got out, only to be thrown onto my face against the curb." In fact, the oscillations were so se-vere that it was impossible to walk. Abandoning his car, which was now sliding from side to side, Coatsworth started to crawl the four hundred yards back to safety. Halfway, he remembered

"Galloping Gertie's" macabre dance of death, as recorded by *Tacoma News Tribune* photographer James Bashford.

he had left his daughter's cocker spaniel, Tubby, in the car. He turned but quickly realized that going back was impossible.

It was all he could do now to save himself from falling off the bridge. Like some real-life fairground ride, it seemed to be reaching angles of thirty-five to forty degrees as it swung violently from side to side. All around him he could hear the sound of concrete cracking. "My breath was coming in gasps; my knees were raw and bleeding, my hands bruised and swollen from gripping the concrete curb," he recalled. Toward the end,

Coatsworth even risked rising to his feet and running a few yards at a time. Staggering and stumbling, he eventually reached the road on the bank and safety.

At one point when things seemed to have quieted down somewhat, Farquharson got onto the bridge and tried to drive Coatsworth's car off. But he soon abandoned the effort. Farquharson himself recorded undulations measuring twenty-eight feet from crest to trough. The edge of the deck was now tilting at an angle of even more than forty-five degrees as the roadway swung from side to side like a hammock in the wind.

This macabre dance of death was recorded for posterity by Farquharson himself—who, incredibly, managed to stay on the roadway near one of two suspension towers to get pictures of the abnormal twisting motions. His motion pictures were complemented by those of another cameraman, Barney Elliot. Both filmed on 16 mm film while *Tacoma News Tribune* photographers James Bashford and Howard Clifford took numerous stills. The astonishing pictures taken by all four men show the violence and direction of the undulations. These were records that would form the centerpiece of the subsequent investigation as well as being the subject of study by thousands of engineering students in the sixty years since.

Shortly after 10 A.M., the final collapse sequence began when the stiffening girders at mid-span buckled. The suspension cables then snapped, snaking up into the air above the main cables. With the center span severed, a section of the roadway at the quarter point now collapsed. Coatsworth's car and Tubby, his daughter's poor cocker spaniel, went with it. There was a brief pause before another six-hundred-foot section of the bridge tore away from the suspensions and plummeted into the icy waters below. More sections of the roadway followed and, as each fell, shock waves rippled along the remaining sections. These shock waves tugged at the side spans and the towers which were pulled backward toward the banks of the Narrows, bending by as much as twelve feet.

As news of the collapse spread, structural engineers registered their shock. "I'm completely at a loss," said Leon Moisseiff. Burt Farquharson was later to note that the failure of the bridge came as such a shock to the engineering profession "that it is surprising to most to learn that failure under the action of the wind was not without precedent." It was a measure of how short memories were; of how few suspension-bridge engineers were rigorously incorporating the lessons of the past into their designs. For few suspension bridges had ever failed but for the "action of the wind."

A COMPLETE INVESTIGATION "OF THE DESIGN, THE BEHAVIOR, AND THE failure" of the Tacoma Narrows Bridge began immediately. The appointed forensic-failure sleuths, a team of three engineers which included Othmar Ammann, designer of the George Washington Bridge, were to report their findings to the Federal Works Administrator, John Carmody. In March 1941, the three engineers handed over a report that concluded the bridge had failed as a result of "dynamic wind-induced vibrations, which had played on its lack of torsional stiffness." The comparatively narrow width gave the bridge adequate lateral rigidity but, combined with extreme vertical flexibility, its narrowness made it highly sensitive to torsional motions created by aerodynamic forces.

Ironically, the investigators concluded that the bridge's final hour was actually a testament to the quality of its design and the materials used in its construction. Before the collapse, the structure and its connections had been subjected to stresses far beyond those it had been designed for. The report emphasized the need for rigidity in two planes—the vertical and the horizontal—and urged that greater attention be paid to dynamic response and structural damping. But the report could not provide all the answers. Indeed, the three investigators admitted that, given current knowledge, they had done little more than identify the problems. "A complete analysis of these [aerodynamic] forces and of the response of a suspension bridge thereto is not possible with present knowledge; further observations, experiments, and mathematical analyses are required," the report concluded.

A spate of research and debate followed. Engineers already knew that stiffness depended on the ratio of the distance between the two suspension towers (L to the engineers) divided by the depth of the stiffening girders across the roadway (D). The higher the ratio, the more flexible the bridge becomes. With a distance between the towers of twenty-eight hundred feet and stiffening girders of just eight feet in depth, the Tacoma Narrows Bridge had a very high span ratio of 1:350. This was in fact the highest L/D ratio for any suspension bridge anywhere in the world. It compared very unfavorably to its closest model, the Bronx–Whitestone Bridge in New York, which had an L/D ratio of 209. The other critical ratio—width of the roadway (W) in relation to the length (L)—was no better at the Tacoma Narrows Bridge. With just two lanes of traffic, a roadway thirty-nine feet wide, it had the highest L/W ratio of any suspension bridge anywhere.

Since 1925 there had been a tendency to construct longer and lighter suspension bridges. Designers were suffering from a number of delusions. Firstly, there was a belief that girder stiffness was not necessary to control the geometry of the roadway. Secondly, there was a belief that the aerodynamic forces that had brought down lighter and shorter suspension bridges in the past would not affect a structure of the magnitude of the Tacoma Narrows Bridge. Its length, after all, did give it more weight, even if it made it more flexible. Thirdly, the forces exerted by the wind were calculated as static, steady, horizontal pressures against the side of the bridge. In reality, the wind exerted aerodynamic forces that were anything but regular in either plane or force.

To understand the effects of aerodynamic forces on the bridge, imagine a hair-drier blowing on a narrow strip of paper. Depending on the angle of the drier, the paper will both bend and twist. The wind, rarely if ever perfectly horizontal, will hit the windward edge of the bridge first, raising or lowering it depending on whether it is blowing from above or below. That in turn will expose the leeway edge, which will move it in the opposite direction. The bridge is literally twisting in the wind.

The Federal Works Agency (FWA) report concluded that even with more research it would never be possible to establish cut-and-dried rules or formulas for the design of long suspension spans; each would, the investigators concluded, require analytical study. Othmar Ammann amplified this view in 1941. He wrote: "The Tacoma Narrows Bridge failure...has shown [that] every new structure that projects into new fields of magnitude involves new problems for the solution of which neither theory nor practical experience furnish an adequate guide. It is then that we must rely largely on judgment and if, as a result, errors or failure occur, we must accept them as a price for human progress."

To many that sounded like a cop-out—an excuse for the profession's failure to do adequate research and experimentation before starting design or construction. The simple truth was that none of the factors that caused the Tacoma Narrows Bridge disaster were unknown. Large suspension bridges had been collapsing under the effect of wind for more than a century. In fact, this had been happening regularly every thirty years or so, in itself a warning to engineers. Thirty years might be considered the complacency wavelength, perhaps.

The same lessons, it seemed, had to be relearned by each new generation of designers. The Tacoma Narrows Bridge was rebuilt, incorporating all the necessary features. It had four lanes of traffic, not two, and stiffening trusses thirty-three, not eight, feet deep, while the whole structure was dampened to counter aerodynamic oscillations. More than fifty years after its inauguration, it is still standing. Hopefully, the dramatic film footage of its predecessor's collapse—still compulsory viewing in engineering classes throughout the world—will be enough to prevent a new generation of bridge designers having to learn old lessons the hard way, through physical failure.

THE OBSERVATION THAT THIRTY YEARS MIGHT BE THE COMPLACENCY wavelength for bridge engineers seems to be borne out by the next major bridge collapse in the United States. The sudden failure of the Point Pleasant Bridge, the U.S. 35 route crossing

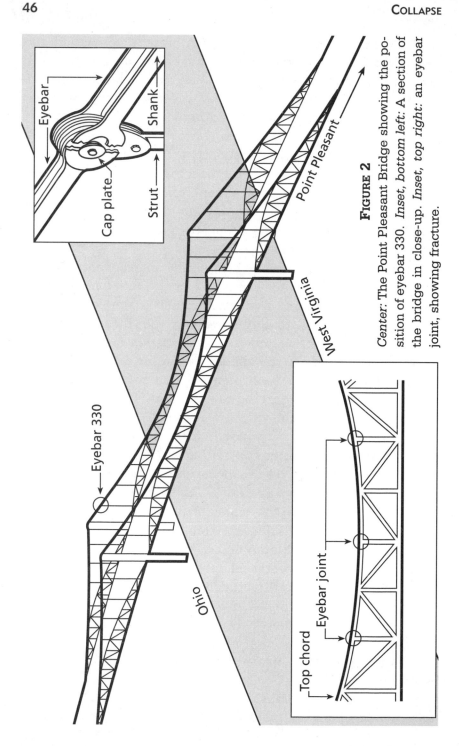

FIGURE 2

Center: The Point Pleasant Bridge showing the position of eyebar 330. *Inset, bottom left:* A section of the bridge in close-up. *Inset, top right:* an eyebar joint, showing fracture.

over the Ohio River, the border between West Virginia and Ohio, was the second most deadly bridge failure ever in the United States. Only the Ashtabula Creek disaster of 1876—when a 150-foot-long cast-iron bridge gave way under the weight of a passenger train, killing ninety-one and injuring sixty-four—was more costly in human terms.

Known locally as the Silver Bridge as a result of the shiny aluminum paint used to prevent its steel components from rusting, the Point Pleasant Bridge had been completed in 1929—a suspension bridge with a double difference. One: it was the first "eyebar" suspension bridge in the country—in other words the roadway hung from two bicycle-type chains made up of fifty-foot-long "eyebars" rather than steel cables. Two: it was the first bridge in the United States to use a recently developed material—high-strength, heat-treated carbon steel. As such, the bridge was the embodiment of risk: it was a new structure, on a new scale, using new materials.

The key feature of the design was the eyebars, untried and untested themselves and composed of a similarly untried and untested steel. Two inches thick and a foot wide, the "eye" at the end of each fifty- to fifty-five-foot bar was connected to the next one in the chain by means of a thirteen-inch pin held in place by a bolted cap plate (see Figure 2, inset top right). There were two eyebars in each link of the chain, road side and river side, so to speak, with the two being connected by the pin. The doubling up was needed: a lot hung on the eyebars, quite literally. Not only were they the suspension element of the bridge laterally, they formed the top of the stiffening trusses of the bridge horizontally (see Figure 2). Hanging down from the middle of the pin, sandwiched between the two eyebars, hung another steel "eye." It formed the top of the vertical steel struts from which the roadway hung.

Just before 5 P.M. on December 15, 1967, the bridge was loaded with traffic—shift workers rushing home, Christmas shoppers laden with gifts and supplies, regular freight trucks. The light was fading and the skies blackening. It looked like snow, Charlene Wood thought as she approached the traffic lights that controlled the flow of traffic on to the two-lane-wide, 1750-foot-long bridge. At the head of the line, Charlene watched the light turn from red to green, engaged her clutch and moved forward to cross to Ohio.

"As I started to go up onto the bridge there was this really terrific shaking. It's something you can't really imagine unless you were there," recalls Wood. At that moment she remembered how her father, a riverboat captain, had always speculated about what would happen if a boat on one of the world's busiest inland waterways hit one of the bridge's piers. "At that moment it clicked and I thought, there's no way I'm going to cross that bridge. I threw my car into reverse, not looking behind me, and as I was backing up the bridge just started folding. It was such a terrific shaking that the car stopped on me but it kept rolling back because it was on an incline. By the time my car did stop, my wheels were right on the very edge of the concrete where it broke off."

Roy Sayre, a truck driver, who lived on the banks of the river on the Ohio side, was in his kitchen cooking supper for his children. "I heard a loud cracking, thundering noise. At first I thought two cars had crashed in front of the house but when I looked out of the window I saw the bridge falling," he recalls. Sayre rushed to the telephone to alert the State police. The call was timed at 4:58 P.M. He had actually witnessed the whole failure sequence, although that in itself was just a matter of seconds. "It was hard to believe and the police didn't believe it either," he muses, recollecting the scene. "It just trembled a bit, then the roadway turned over and fell out. You could see the cars fall and the floor—slow motion—just fell right on top of them."

Charlene Wood recalls: "The bridge just started folding. It looked like dominoes...dominoes all lined up, and you know how you give them a push and they start coming forward at you. There was a car ahead of me that went down but I have no idea who it was....You could hear the debris hitting the water, the splash came up, you could see it." Safely reversed up route U.S. 35 on the West Virginia side of the bridge,

Wood's car was greeted by a State patrolman running up the road toward the disaster. He opened the car door and helped the heavily pregnant Charlene out. "You could hear screaming and hollering," she remembers. "It was the people who were in the water."

Within minutes Wood was in an ambulance, on her way to Pleasant Valley hospital. Her quick thinking—brought on by an awareness of what might be happening—had saved her life but now she had to be treated for shock. "I'd kept cool until it was all over but then I just lost it. Everything was fuzzy. I didn't want to think what had happened." Within two months she had given birth to twins. But the memory remains; the fear of crossing bridges never goes. "I just have to bear with what happened. I thanked God...but I just wondered why I was spared. It bothered me."

Roy Sayre ran down to the river bank. His wife, who might have been on the bridge returning home, returned safely but it was quickly clear that many other families would not be so

The Point Pleasant Bridge—known locally as the Silver Bridge—over the Ohio River. completed in 1929, it was the first eyebar suspension bridge in the United States.

lucky. Using the headlights from his truck—a truck in which Roy himself had crossed the bridge earlier that day—rescuers could see cars crushed beneath the twisted mass of concrete and steel that had plummeted into the Ohio River. Boats arrived to pick up survivors in the water but before long it was realized that cutting and heavy lifting equipment would be needed to recover many of the thirty-one vehicles that had fallen into the icy waters.

Murky water up to seventy feet deep, rain and snow, the pitch dark, a six-knot current, and the fact that the layers of debris were now blocking a busy shipping lane all made life difficult for the rescuers. Of the sixty-four people who had hit the water, only eighteen were eventually picked up alive. By daylight, cranes overhead and divers down below were uncovering flattened vehicles and the bodies inside. One of the forty-six victims, a taxi-cab passenger, was recovered clutching the dollar bills that were to have been his fare—evidence of the speed with which it had all happened.

AS THE SEVERED PIECES OF THE BRIDGE WERE COLLECTED AND DUMPED IN a twenty-seven-acre field on the banks of the river, the scale of the painstaking task of finding a cause became obvious. Eliminating the usual suspects—overloading, age, fatigue, an impact, aerodynamic instability, and corrosion—was going to be complicated, not least because so much of the critical evidence had been interfered with: cut by the rescuers, rusted by the river and rain, dumped on the river bank without regard to sequence or order. "I've never seen such a pile of junk," complained Ed Young, associate editor of *Engineering News-Record,* at the scene. "Every joint is broken and every broken piece is rusted. It's going to be almost impossible to separate old breaks from new."

When John Bennett of the U.S. Bureau of Standards arrived to investigate the cause he agreed. "The Ohio River there is very heavily traveled so the U.S. Corps of Engineers had taken all the debris and had just piled it on the shore—it was a terrible mess," he recalls. "Fortunately, each piece had been photographed as they took it out." Leafing through the photographs, Bennett soon alighted on one of particular interest. It was a picture of an eyebar that had sheared in half—fractured through both sides of the hole that was the eye, rather than through the shank. "Looking at it, the fractures on the two sides were completely different. On one side it was very straight, almost like a saw cut. . . . The other side was extensively deformed, the metal bent and the paint chipped off."

The first thing the sheared eyebar suggested was that the bridge had not failed as a result of overloading. If there had been too much weight, it would have failed in the shank. The crucial photograph seemed to eliminate a major possible cause. The massive rise in traffic meant that live loads on interstate highway bridges had increased hugely since most of them were designed. The Point Pleasant Bridge was forty years old when it collapsed and for years had been shouldering more than its design weight. However, John Bennett's observation matched that of the recovery engineers. Thirty-one vehicles on the bridge at the time of its collapse hardly constituted overloading.

John Bennett was convinced that the eyebar was the initiator of the collapse even if other factors were involved. There was one problem. The outer part of the eyebar had broken off and presumably fallen into the river. It had not been recovered. He sent the army divers back into the water to trawl the mud and rock of the riverbed in the area it had hung in—upstream, the second eyebar west of the Ohio-side tower. Eventually they found the remains of what had now been identified as eyebar 330. But that was not John Bennett's only piece of luck. Apart from the fracture surface, the remains of eyebar 330 were completely undamaged—it had apparently fallen into the river without hitting anything on the way down.

The piece was sent to Bennett's laboratory for examination. It had a light coating of rust on it from its time in the river. He handed it to a technician who, on looking at the fracture plane, immediately noticed something. "He said: 'John, look at this.' In one corner of the fracture surface there was an area covered with a dark coating, completely different from the light rust

The tiny area of deeply encrusted rust discovered inside the metal of eyebar 330. It indicated that a fatal crack had developed during the forging of the steel forty years before the bridge collapsed.

that covered the rest of the surface," recalls Bennett. It was a tiny area of deeply encrusted rust, no more than an eighth of an inch long (see photo, above). But it could have been enough to play a vital part in the failure sequence. It indicated there had been a crack at that location prior to the final fracture, a crack that might—given the slow speed with which high-strength steel rusts—go back as far as the forging of the steel forty years earlier.

This all fit. Tests had confirmed what the investigators had believed from the appearance of the fracture in the first eyebar: this was a brittle, sudden fracture, the sort you get in breaking glass. But the high-strength steel from which the eyebars had been forged had plenty of ductility—that was one of its main properties. Indeed, the companion eyebar, the one opposite 330 on the other side of the bridge, had shown precisely these properties. The edge of its eye was distorted—rolled back and flattened where it had obviously been under extreme pressure—as it took the full load of the bridge for some seconds after its companion had fractured.

It was well known in the field of fracture mechanics that the most common cause of brittle failure in elastic material was the concentration of stress at the site of tiny flaws or cracks. Examining cross-sections of the metal eyebar under a microscope, investigators found what seemed to fit the bill: a number of minute cracks as small as pinholes, apparently formed when the steel was forged back in 1926. One of these, it seemed, had grown into a crack inside the eyebar, invisible to inspectors or maintenance workers. This crack, the cause of the internal rust observed by the Bureau of Standards technician on the outer part of the eyebar recovered from the river, was just an eighth of an inch deep and a quarter of an inch long at the time of the disaster.

Over the years the crack had grown, slowly but fast enough to be potentially fatal. Its silent, invisible, almost imperceptible growth was the result of what structural engineers, now in 1968, forty years after its construction, understood as stress corrosion—the combined effect of the application of a tensile stress and the sort of pollutants that can eat away at metal, causing corrosion. Although the investigators were never able to pinpoint what had had most effect there were plenty of candidates in and around the Point Pleasant Bridge—chemical plants, coal-burning locomotives traversing the nearby railway bridge, emissions from the traffic that crossed every day.

Corrosion had always been a major problem here—hence the anti-rust silver preservative that gave the bridge its nickname. But a feature of the design of the eyebars made things much worse. The eye of the eyebar was a little larger than the pin. This was simply for ease of assembly and in physical terms was nothing—just one-eighth of an inch. However, the gap created an airspace into which pollutants could easily enter, where corrosion could develop unabated and undetected. Indeed, this gap was the worst of all possible worlds. It was not big enough for inspectors or maintenance staff to be able to see into and check, yet was large enough for atmospheric pollution to get in to corrode.

But the stress corrosion inside the eyebar might not have mattered if the steel used in the bridge had not been high-strength. "The higher the strength, the more brittle the steel can become," says John Smith, an expert metallurgist at the National Institute of Standards and Technology. And one key determinant of high-strength steel's brittleness is temperature.

"Steel like this is much more brittle at low temperatures and much more ductile at higher temperatures. In this particular case it was a relatively cold day, just about freezing." As both Roy Sayre and Charlene Wood had recalled, it looked as though it might snow.

Ironically, investigators were able to work toward answering the question of what had gone wrong at Point Pleasant Bridge only because of the huge advances made in fracture mechanics in metal since the Second World War. The main stimulus to this research had been the loss of several welded high-strength steel ships in the Arctic Ocean in the early 1940s. These Liberty Ships, conveying military supplies to the Russians, had been lost after literally cracking up in the freezing cold, according to survivors. They too, it seemed, were victims of a combination of tiny cracks and the brittleness of such steel in low temperatures.

New research was essential. With the increasingly widespread use of high-strength steel in construction and shipbuilding, structural engineers needed to know what caused a crack to propagate and how such propagation could be resisted. It was known that temperature was crucial, but what temperature had what effect? A test known as the Charpy V-notch test was soon developed to quantify the brittleness of metals at different temperatures. A pendulum was swung to hit a metal bar notched to induce a stress concentration. Greater force was steadily applied by swinging the pendulum from steadily greater heights.

Variations on this test—in particular reproduction of the stress levels inside the eyebar—were now applied to samples from eyebar 330 at the Battelle Memorial Institute in nearby Columbus, Ohio. The results made it plain what difference temperature made. At 32 degrees Fahrenheit, the temperature at the time of the disaster, the fracture energy required to break the samples taken from the eyebar was 2.2 foot pounds for the central layer of steel and 2.6 foot pounds for the outer layer. At 165 degrees the force required was about triple this: 6.8 and 8.6 foot pounds respectively.

"Had it been warmer, say a summer temperature of seventy-five degrees, that particular steel would not have failed unless the initiating crack was perhaps half an inch deep and an inch long," concludes John Smith. "That size crack might have taken a thousand years to grow to those proportions." But at about freezing, the minute crack in eyebar 330 was enough. "We asked the metallurgists at Battelle how big a crack was necessary to cause rapid crack propagation in a particular material at a particular temperature under a particular stress," says John Bennett. "All their tests indicated that given those conditions a crack one-eighth of an inch deep would be sufficient."

But how had the pinhole-sized cracks formed in the first place? Were there more lessons to be learned here? Given the presence of others clearly visible under the microscope, clusters of pinprick-size pits in the core of the eyebar, the investigators decided to home in on the steel-forging process more than forty years before the failure. They discovered that these flaws were actually the product of a quenching and tempering process designed to minimize the brittleness of the steel to make it more ductile and thus more resistant to stress.

"Quenching and tempering involved heating the eyebars to about fifteen hundred degrees, plunging them into water to cool them down, then tempering them," says John Bennett. "Now, obviously if you heat something to fifteen hundred degrees, then cool it in water, the outside is going to cool a lot faster than the inside." The result was a softer, more ductile metal on the surface, but a harder, more brittle layer below. It was in this slower-cooling area beneath the surface that the pinhole-size flaws appeared. "In the late 1920s they just didn't know that the relatively slow cooling which occurs in a two-inch-thick piece of steel results in a material that has quite good strength under a static load but is sensitive to crack propagation, very sensitive to brittle fracture."

It was easy to see how the minute crack, covered by the pin and the cap plate, would have been impossible for maintenance engineers and inspectors to spot. Less obvious was why the failure of one eyebar would have resulted in the rapid collapse of the whole bridge. Failure of a single wire in the steel cables more commonly used in suspension bridges never resulted in such spectacular failure. Indeed, the wild oscillations that led to the collapse of the Tacoma Narrows Bridge wore or broke five hundred of the eighty-seven hundred wires that made up the cables from which the bridge hung. Yet while the roadway fell into the water, the cables remained in place.

The answer lay in the design which was completely "nonredundant." The way in which the struts, pins, and eyebars were all so interdependent meant that total collapse was almost a built-in design feature. Any one failure meant that everything failed: any individual failure in the chain could literally set off a chain reaction, which it did. Once eyebar 330 had gone, the stress would be passed on to its opposite number on the other side of the roadway. There had been no allowance for such secondary stresses in the design. With one eyebar gone, the Point Pleasant Bridge was a disaster waiting to happen.

Ironically, the design loads had been radically reduced as a result of the choice of design. When J. E. Greiner, Consulting Engineers of Baltimore, Maryland, prepared plans and specifications for a suspension bridge over the Ohio River at Point Pleasant in May 1927, the original project specifications listed strength and stress levels double those eventually adopted. Had the bridge been a conventional cable suspension bridge, the steel in the cables would have had to have a minimum ultimate strength of 220,000 pounds per square inch (psi) and a yield strength of 140,000 psi, and an allowable working stress for the same steel of 80,000 psi. When the chain-link design submitted by the American Bridge Company was accepted, the ultimate strength of its heat-treated steel was 100,000 psi with a yield strength of 75,000 psi and an allowable working stress of just 50,000 psi.

Incredibly, despite the uncertainties of the unproven design, the lack of protective features in the eyebars, and the lack of structural redundancy in the whole bridge, the safety factors built into the bridge had been diminished rather than improved. But there was something else. Even before the construction of the Point Pleasant Bridge someone had raised the alarm about the nature of the high-strength steel being used in the construction. They had also raised concerns about the American Bridge Company's refusal to make public its testing or processing of the material.

In 1927 there was only one other chain suspension bridge using heat-treated eyebars in existence. It was in Florianopolis in Brazil; it had been completed in 1924–25, constructed by the American Bridge Company, who had patented the heat-treated eyebars. The designer of the Florianopolis bridge, D. B. Steinman, had had reservations about some aspects of the design and had introduced a number of extra safety measures, in particular thickening the heads of the eyebars and doubling the number of eyebars in each link of the chain from two to four. Both these precautions were ditched in the Point Pleasant design.

But it was something else that most preoccupied Steinman. In 1924, three years before the design for the Point Pleasant Bridge was accepted and forty-three years before it collapsed, Steinman had complained that the material and processing of the steel being used in these chain-link suspension bridges had not been subject to any independent testing or scrutiny. The formula and contents of the steel and the heat-treating process it was being submitted to were being kept secret from everyone—even the consulting engineers, Steinman complained. Writing somewhat prophetically in *Engineering News-Record* in November 1924, Steinman stated: "Under those circumstances the consulting engineers declined to assume any responsibility for the strength and safety of the eyebars furnished for the structure and the matter was left with the understanding that the American Bridge Company assumed complete responsibility for that material."

In the end, there was no one to assume responsibility. As so often, the absence of independent external scrutiny of internal company results masked the fact that the key tests on the eyebars had not actually been done. In 1968 Bill Thatcher, then assistant to the head of engineering at American Bridge, told accident investigators that the company had no record of ever having tested the fifty- to fifty-five-foot eyebars used in the bridge. In fact, the testing equipment could not handle this length of eyebar. As Steinman put it when quoted in an article marking the twenty-fifth anniversary of the disaster: "The testing was relegated to special pieces of heat-treated steel manufactured for the sole purpose of testing."

Remedial measures did follow the collapse of the Point Pleasant Bridge. A twin bridge of the same design also over the Ohio River, at St. Mary's, West Virginia, was dismantled; President Lyndon Johnson ordered an inquiry into bridge safety throughout the nation. New tools such as ultrasonic probes and acoustic emissions devices were developed to aid inspections. A survey of individual states indicated just how haphazard bridge inspection and maintenance was, a fact borne out by the record in the Point Pleasant case. West Virginia's State Road Commission bridge inspector Paul McDowell had admitted during the investigation that he had not followed nearly half the bridge inspection procedures outlined in his inspection manual when he had last surveyed the bridge in 1965.

More than seven hundred thousand bridges were inspected and classified by Lyndon Johnson's panel on bridge safety. A national policy for bridge inspection became law when Congress approved the National Bridge Inspection Standards Act (NBIS) in 1968 and a massive inspection process began as the funds were appropriated. Thousands of defects were detected in hundreds of bridges. A new bridge replaced the Silver Bridge in 1969. It was not a chain-link suspension bridge and it was properly inspected. Indeed, in 1977 a routine inspection found a crack in a weld of a steel girder. The bridge was closed for repairs to sixteen welds. In this case, at least, the new system was working.

3

DESIGNING, DIGGING, DISASTER

Abbeystead

SHEEP, ROLLING HILLS, DRYSTONE WALLS, GURGLING STREAMS—THE rural delights of the valleys and fells of upper Lancashire, England, are an unlikely location for one of Britain's most baffling structural engineering disasters. But it is not just the setting on the edge of the Forest of Bowland that is so incongruous. A construction, yes, but one that was underground; an explosion, yes, but miles from any of the most likely industrial causes; a design disaster, yes, but what could go wrong with the design of a simple pumping station?

The geography here is the key. Abbeystead Vale lies between two rivers: the Lune, a fast-running river draining water from the edge of the Lake District into Morecambe Bay, and the much smaller Wyre, eight miles south. Demand for water in the industrial towns to the south—Blackburn, Wigan, and Preston—all of which could be supplied from the Wyre, was growing. The obvious solution was another reservoir. But it had been difficult to get planning permission in the areas of outstanding natural beauty that were the obvious catchment areas for the water. It had been even more difficult to sell such schemes to the public.

For the water authorities there was another possibility. It had the double advantage of being relatively cheap and probably saleable to both the politicians and the public: find a way to pump water from the lower reaches of the Lune to the upper reaches of the Wyre. The water company already had extraction facilities on the lower Wyre near the town of Garstang. The new scheme would allow it to take more water at this plant while minimizing the environmental impact. There was nothing unusual in this. The area was already laced with pipelines, water tunnels, and boreholes—the Whit Moor borehole and Bowland Tunnel, to name just two.

The only obvious problem was Lee Fell, a 376-foot peak that blocked the way three miles into the route. The only obvious solution was an inclined tunnel through the hill. So in 1970 a scheme was drawn up as part of what was termed the Lancashire Conjunctive Use Scheme. It allowed the water authority to take up to seventy-four million gallons a day from the Lune above Forge Weir near Skerton, provided the flow at the weir was at least ninety-six million gallons a day. Screened for debris and fish, the water was to flow to a pumping station by gravity, then via the first of two settling tanks to a balancing tank with a capacity of nearly two million gallons—a constant head of water for pumping. A second set of pumps would then push the water through 3.35 miles of pipeline and 4.1 miles of tunnel running from the Rowton portal to the Abbeystead Valve House outfall where the water would enter the River Wyre (see Figure 3).

Design and preparatory work went on from 1971 with public consultations leading to a number of changes. Plans to chlorinate the water passing between the two rivers were dropped after objections from anglers. Questions were raised about the security of the system, the landscape blight, and protection of the pumping valves from the freezing winter temperatures on the fells. The answer to all three concerns, the planners decided, was to build the proposed valve house at Abbeystead underground. The plans were adapted accordingly and no major objections were raised at a public inquiry held in Lancaster, England, in March 1974. Planning approval was granted the following year.

Edmund Nuttall Ltd. soon began work on designs drawn up by Binnie and Partners. The major challenge for the construction team was the tunnel. Information on the geology of the area was limited—the standard Ordnance Survey geological maps were based on surveys done a hundred years earlier. But the rock did not look too difficult: upper carboniferous strata of the millstone group, as the geologists classified it. There were in fact no real problems. The tunnel was bored from both ends, initially with machines and shield drive, later by drilling and blasting. A few boreholes were drilled to get additional data on the rock but these were limited to the areas where there would be little ground cover for the tunnel—the constructors' main concern.

Such problems as there were during tunneling were not unusual. Traces of flammable gases were detected and there was some inflow of water through the tunnel walls. However, most of the tunnel was not designed to be watertight. Apart from 437 yards at the Rowton end and another 656 yards at the Abbeystead end, both of which were lined with steel to withstand the pressure of pumping, the tunnel, with a diameter of nearly nine feet, was nothing more than porous concrete. Relative pressure inside and outside the tunnel was the key to how much water would actually enter or exit it. Water ingress was initially estimated at about forty-two pints a second. However, on completion the estimates were lowered markedly to about twenty-five pints a second when the tunnel was empty and seventeen pints a second when the tunnel was full, a total of about 264,200 gallons a day.

Construction work finished in the spring of 1979 and water was first pumped through the system in June. It was consid-

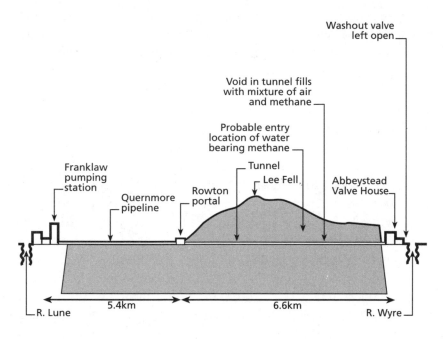

FIGURE 3
Cross-section of the Lune-Wyre Transfer link.

ered an important new development, important enough to be opened by Queen Elizabeth in 1980. Yet ironically it was markedly underused. The pipeline could handle seventy-four million gallons per day yet no more than twenty-three· million gallons was ever pumped. On average, the pipeline was only used every other day.

However, local people downstream from the Abbeystead end of the link were convinced the Lune-Wyre Transfer link was having a major impact on their environment. On completion, the river seemed to flood very easily. In October 1982 there was serious flooding after four or five days of solid rain. The river burst its banks and flooded the tiny village of St. Michael's-on-Wyre, some fifteen miles downstream from the Abbeystead outflow. George Tyson, a local farmer, woke up one morning to find all his land under water, and his father's bungalow flooded through with more than three feet of water. "We thought the transfer of water was probably one of the major reasons the Wyre came up from absolutely nothing to full—dangerously full—in a very short period of time," says Colin Burke, a local garage mechanic.

In the spring of 1984 it happened again—the worst floods in fifty years. But it was not just the scale of the flooding, it was the direction. The Wyre burst its banks at Churchtown, a village upstream, and St. Michael's was flooded overland from behind. "The flooding was very unusual...the water coming from behind us, from Churchtown. I couldn't follow that," recalls Colin Burke. As the village flooded again it became accessible only by boat or tractor. Colin Burke was lucky—he had a boat. Walking through the tiny village now, he demonstrates the scale of the disaster by pointing to a three-foot high wall. "A good indication of how high the water was is that wall at Mr. and Mrs. Tyson's bungalow—we sailed over their wall."

The matter of the impact of the Lune–Wyre Transfer Link was raised with George Tyson as chairman of the parish council; he in turn raised villagers' worries with the North West Water Authority (NWWA). A visit to the Abbeystead Valve House and the pumping station at Franklaw was suggested as the best means of alleviating fears. Eventually a date was set—May 23, 1984. Having canvassed for interest, George Tyson decided there would not be enough people on the trip to hire a bus. "George rang me at work that day and said they were short of cars, could I take mine," says Colin Burke. Colin rushed home after work, then drove to the meeting place, the pub. His wife, Denise, decided not to go, realizing their one-year-old baby's car seat would take up the back seat that could otherwise transport passengers.

The group from the village numbered thirty-six in all, traveling in eight or nine cars. It was a beautiful evening for the fifteen mile drive north to Abbeystead, up the A586 highway through Churchtown and Garstang, across the M6 motorway, and into the fells of the Forest of Bowland. Indeed, it seemed an incongruous time to be investigating flooding. The spring downpour had turned into a summer heatwave. St. Michael'son-Wyre had just enjoyed more than a fortnight of gorgeous Mediterranean weather. "We were going for a pleasant evening out, a village outing as it were. There were people of all ages there, from young Mark [Eckersley] to some of the grandparents," recalls Colin Burke. "We were simply going to a waterworks where they would show us that their water system could not possibly flood our river. It was just a very relaxed evening."

Joined by eight staff members from the North West Water Authority, the proceedings started when the villagers sat on the grass outside the pumping station for a brief introduction to the water link and the role of the pumping station. The structure they faced, known as the Valve House, looked something like an underground concrete bunker built into a man-made landscape, with a single exposed wall for access and soil and grass covering the roof. Entrance was through outward-opening metal double doors, fifty yards from the river and the downstream end of the outfall.

As the party crossed the threshold they crowded into what was known as the dry room, forty-four of them in a space forty-one feet long by sixteen feet wide. The dry room was directly above the discharge branch of the tunnel, housing the instruments, controls for the water valves, a toilet, and a small management office. A second pair of substantial double doors in an internal dividing wall opened from the dry room into what was known as the wet room. This was shorter but wider, thirty-one feet by twenty-five feet, with huge concrete beams and neon lighting overhead. The wet room occupied the whole area above the water discharge and distribution chambers.

As such, the room's main feature was its floor, which was essentially water. The visitors now found themselves walking on huge steel meshes carried on joists more than twenty-six feet above the bottom of the discharge chambers. "It was like being on top of a big well," recalls George Tyson. "We just walked about on this grid looking at the water." Colin Burke remembers the main points of the guided tour. "The idea was that the water came into that Valve House, was distributed among the tanks which slowed it down, controlled it, then passed it into the river through a number of pipes on a stone bed. We were to see the water coming in and going out—see that as it was discharged it couldn't flood the river, that it was all controlled."

The water in the discharge chambers was only two to three feet deep so water would have to be pumped from the Lune pumping station to demonstrate how the link worked. Speaking on a telephone from the dry room, Alan Lacey, the NWWA District Supply and Treatment Manager, instructed the treatment controller at the Lune end to start pumping. At 7:12 P.M. a pump that would push water from the balancing tank at the Lune end into the tunnel was switched on. It would be the first time water had been pumped through the tunnel for seventeen days.

Colin Burke remembers asking how long it would take for the water to come. "Mr. Lacey said he didn't know because he had never been there before when the pumps were switched on. He made quite a fuss telling us how they'd borrowed the water for this demonstration and how it would all have to be given back." Expecting the water to arrive any minute, some of the party left the Valve House to walk the fifty yards down to the river to see the water pour out of the outfalls on both sides of the river bank. There was little more than a trickle emerging.

When the rush of water did not arrive after about ten minutes, Alan Lacey made a second phone call. He asked for a second variable speed pump to be started up, effectively doubling the rate of anticipated flow. By then, virtually all the visiting party were back in the Valve House wondering what the delay was. After twenty minutes, some of the visitors were beginning to think there might be a serious problem; others began to think it was all looking a bit like a sham. Why had NWWA offered a demonstration they couldn't give? Why hadn't the equipment been tested before they arrived?

For many these were their last thoughts. Tim Eckersley was staring down through the grid. Mark, his eight-year-old son, wanted to know if there were any fish. Tim remembers the whole of the air beneath the grid turning a brilliant blue. George Tyson had been outside watching the river when, tired of waiting, he came inside to find Colin Burke. The only sensation he remembers is that of an intense heat totally enveloping him. Colin Burke himself had been standing with a group of friends talking to some NWWA officials. He remembers the whole place lighting up blue as he was blown through the air.

The massive explosion lifted the concrete-beam roof into the air before bringing it crashing down into the Valve House, crushing its occupants. The collapsing beams and debris broke through the two-inch-thick steel mesh of the wet room floor, taking several of the visiting party down into the water tanks. Many were unconscious; others were screaming and moaning. Colin Burke found himself lying in a pile of rubble at the main entrance. As he came to, he saw George Tyson walking around somewhat dazed, his clothes all burnt off save his underpants and the belt of his trousers. Colin raised his arms to try and extract himself and realized that there was no skin left on his hands.

But George looked as though he might be in an even worse way. "As he passed me I could see the skin off his back was hanging over his bottom in a sheet," says Colin Burke. "The whole of his back was just like a piece of red meat in a butcher's shop." But Colin himself could not even get up. As George and others tried to pull him back from under the lintel of the main doorway in case of further collapse, he noticed his feet were back-to-front, his toes down and his heels up. He gradually became aware of a searing pain in his back. He could not move his legs.

Within minutes the water arrived, flooding what now looked like a bomb site. The victims in the tanks, injured and trapped by rubble, were now in serious danger of drowning. When the pump operator at the Lune Pumping Station rang to check that the water was coming through, one of the first of the local people to arrive—alerted by the noise of the explosion—took the call. On being told there had been an explosion, the pump operator suspected a hoax. But he did ring the main control room at Franklaw Water Treatment Works. About fifteen minutes later the water was switched off.

Ambulances began arriving from about 7:48 P.M., followed quickly by the police and fire brigades. Everything had to be assessed at once. What were they dealing with? How many casualties were there? Was there still a real danger from the cause of the explosion? It soon became clear that eight people were dead. They included Tim Eckersley's son, Mark, and Alan Lacey, the NWWA official, both crushed under the roof. But there were many more that the emergency services thought they could do little for. The severe burns, crushing injuries, and bleeding that everyone at the scene suffered would claim another eight lives before long. The final toll would be thirteen of those who had set out on the village outing from St. Michael's-on-Wyre and three of the NWWA officials.

At the scene, some seemed neither dead nor alive. Colin Burke was one of them. Two badly broken legs; a skull fractured back and front; five vertebrae ripped out of his back; terrible burns, including internal burns of the lungs, throat, and windpipe...it was a long list. The firemen had assumed he was beyond help; hospital staff barely disagreed. When his wife, Denise, arrived at the hospital, having heard about the disaster on the BBC's *Nine O'Clock News,* the best prognosis was that Colin would be in a wheelchair for life. As Colin recalls: "They told her: 'You go home now, dear, and if he's alive in the morning he'll be all right.' So she had to come home to a one-year-old child and wonder whether she dare ring up or go into the hospital in the morning."

Denise did go in the next morning and Colin was still alive. Operations, splints, bags on hands...he spent sixteen weeks in bed before moving to a wheelchair then on to crutches. After nearly two years he was able to walk unassisted. George Tyson was a little luckier. In the hospital for seven weeks, he was off work for about a year, and returned to the farm in the late spring of 1985. He took up the silage-making he had had to abandon to friends and neighbors the previous year.

AS BODIES HEALED AND MINDS FOCUSED, THE CAUSE OF THE DISASTER BE-
came the major point of interest for those hospitalized. Officials
from the NWWA and the Health and Safety Executive (HSE)
made their first trip to the site the following day. Everything
pointed to an air-gas explosion. One by one the possibilities—
leakage from British Gas pipelines in the area, storage of gas
cylinders on the premises—were eliminated. Investigators suspected methane, the gas used for domestic cooking and heating. Testing with a methanometer soon proved them right. The
gas most feared by miners, the gas traditionally detected by
canaries, the gas responsible for thousands of deaths underground, had claimed another sixteen lives at the Abbeystead
Valve House.

A small spark or light anywhere in the Valve House would
have been enough to ignite the highly flammable gas. And the
gas could only have come from the tunnel. But how had it got
there? Why had it collected in such lethal quantities? Why had
it not been detected?

Investigators began a series of tests, examinations, and
forensic analyses. They found a number of factors had led to the
disaster: the geological conditions, the lack of use of the
pipeline for seventeen days, the design of both the pipeline and
the Valve House, and a lack of awareness of all this on the part
of both officials and visitors—everything combining in a lethal
cocktail. The investigators then went on to develop theories
that would be tested in an effort to simulate the conditions in
the pipeline and the Abbeystead Valve House on the night of
the explosion.

Firstly, the tunnel. It had been deliberately left porous; not to
let water leak out but to let it leak in. The reason was simple.
Ground water was plentiful in the area and, when water was
not flowing through the tunnel, the greater ground-water pressure outside compared to the tunnel-water pressure inside
meant useful leakage in rather than out. The inflow through the
cracks and minute holes of the concrete-lined tunnel had been
measured at more than twenty-eight pints per second when
the tunnel was completed; that was 264,200 gallons a day,

The North West Water Authority (NWWA) Valve House at Abbeystead in Lancashire the day after the explosion, its roof and doors blown off.

enough to supply a town of ten thousand people with all the water it needed. When investigators checked this after the accident the ground-water flow was very similar—twenty-three to twenty-five pints per second. Above all, the inward flow meant the tunnel was designed to remain full of water during "standstill" periods when the pumps were not working.

But as investigators walked the tunnel after the disaster, they found clear evidence that it had frequently been left at least partly empty. There were assorted tide-marks—deposits of minerals and dirt—down the inside of the tunnel at the Abbeystead end. But even more curious were the large stalactites attached to the sides of the tunnel at levels that should clearly have been covered by water if it had been left full. Inquiries with specialists confirmed that these would almost certainly not have formed under water. And they would have taken some time to grow. There had been a "void"—as investigators termed it—in the tunnel for some time.

But if the tunnel was empty, how was the water escaping? Following the tunnel back, investigators focused on a washout valve—a simple tap—at the end of the tunnel where the water poured into the Wyre, fifty yards from the Valve House at Abbeystead. This was supposed to be closed. It was not. It was open at the time of the explosion and had been left permanently open, in apparent violation of NWWA's Manual of Operating Instructions. Inquiries revealed that within months of the tunnel opening, operating staff had abandoned the practice of opening the washout valve once a month to flush out "dead water" from the system as laid down in the Instructions. Too much silt seemed to accumulate when it was closed, discoloring the water in the river when the valve was opened. Anglers, among others, had complained. It was left permanently "cracked open"—up to one turn from the fully shut position.

This change in procedure had never been authorized by senior staff—but then it did not seem to need to be. The Health and Safety Executive concluded that the operating procedures were not strict enough. Nor, it seemed, was the inspection system, which had gradually become less rigorous. The change in procedure on the washout valve had coincided with the move to operate Abbeystead remotely from the Franklaw Water Treatment Works downstream. A routine weekly visit by a staff operator had initially included a check on the water flow over the weirs in the wet room—the best indication that the tunnel was full. But this too had been dropped. The operator had in fact made his weekly visit to Abbeystead Valve House just hours before the explosion occurred on May 23. He did not even enter the wet room.

No one had any idea that an empty tunnel could be a lethal tunnel. If water was let out, gas could get in. But from where?

There were three possible sources. The gas could have entered from the River Lune; it might have formed inside the tunnel, the product of putrefying vegetation; or it could have leaked into the tunnel in the same way the water did, through the cracks and porous pits of the concrete lining. It was possible that the origin was a combination of all three sources, but broadly speaking investigators wanted to know if the gas was geological or biological.

The second possibility seemed unlikely after scientists at the Rowett Research Institute in Aberdeen, Scotland, analyzed samples of wet slime on the bottom and walls of the tunnel. They found methanogenic organisms, but these were unlikely to have been enough to cause the explosion. Samples of the gas found in the tunnel were then sent to the Isotope Measurements Laboratory at Harwell in Oxfordshire, England, for radioisotope dating. The methane was at least twenty thousand years old. Its main, if not sole, source was underground. And the highest concentrations of it were found a mile back up the tunnel from the Abbeystead Valve House.

But methane, even if present, is not soluble in water, as the engineers who had overseen construction of the tunnel knew. They were right—but in one crucial respect they were also wrong. Methane is indeed insoluble in water, but only at normal temperatures and pressures. And the pressure in the rocks surrounding the tunnel was anything but normal. The pressure was what kept water flowing into the tunnel in the first place. So on one side of the tunnel methane was quite soluble in water, but on the other, where the pressure was lower, it was insoluble. So having entered the tunnel in the water, the methane would return to a gaseous form. All it had to do was bubble to the surface, collect, and concentrate—to the 5 percent by volume in air at which methane becomes flammable.

On the evening of the disaster the pumps would push the collected methane toward the victims in the Abbeystead Valve House with fatal efficiency. And here the tragedy was compounded by one more unfortunate factor: design. There was no ventilation in the tunnel and a lethal form of ventilation in the

Valve House. The tunnel was the first long water conduit in Britain without ventilation to allow gas to escape. The Valve House did have a vent from the tunnel—in fact, it had a whole vent chamber separate from the house itself. Yet, incredibly, the chamber's duct expelled into the Valve House rather than into the atmosphere. Underground, with a solid lid of concrete, earth, and stone, the Valve House was a pressure cooker—a pressure cooker with no air vent to act as a valve. Contained and constrained, the force of any methane explosion here would be magnified many times.

In the end, everything went back to the methane. If the designers, Binnie and Partners, constructors Edmund Nuttall Ltd., and the operators, NWWA, had believed there was a risk of significant quantities of methane being present and of it being soluble in water, the design and operation of the Link might have been different. As the HSE wrote in the conclusion to its report: "Both Binnie and Partners and the NWWA believed and still believe that methane was not emergent from the strata in quantities which were significant."

In the face of the HSE's findings, neither party saw any issue in their failure to research the geology of the area more thoroughly. Both organizations told the HSE that it would not have been economically feasible to drill more than a "handful of boreholes" to investigate conditions along the length of the tunnel route. Even if they had been warned about methane, they told investigators, their experience during tunneling would have dispelled fears. They would not have taken any measures to prevent methane explosions.

On the face of it there may have been sufficient evidence for prosecution in the HSE's report. Indeed, John Rimmington, the HSE's Director General, stated that the NWWA had to face up to its responsibilities "morally, humanly, and civilly." But such responsibilities did not apparently include legal ones.

With the HSE declining to prosecute, the survivors set out to do their own investigation as part of a possible civil legal action to prove liability. In 1987, Tim Eckersley and the other victims won their case against NWWA, the operators, Edmund

Nuttall Ltd., the constructors, and Binnie and Partners, the designers, all of whom the High Court found negligent and thus liable. On appeal, Binnie and Partners were found liable and ordered to pay the whole of the £2.2 million compensation for the deaths and injuries.

The survivors and relatives of the victims began their investigation by hiring Iain Williamson, an experienced British geological engineer. The first thing he decided to do was travel to Lancashire and walk the length of the tunnel. "I wanted to put myself in the position of being a geologist retained by the consultants designing the tunnel, just to get a feel of the area, get a general view," he says.

Equipped with walking stick and boots, Williamson took a television crew back to the key sites on his walk. Through the Condor valley and the village of Quernmore, Williamson took the crew along what is known as the Quernmore pipeline, the three miles of the link that was not underground. Stopping briefly to point out the peak of Clougha Pike, behind which lies Lee Fell, he marched on to what is known as Rowton portal, the northern end of four miles of tunnel that lead directly to the Abbeystead Valve House.

"I looked around the area of the tunnel entrance here, and there was a lot of disturbed ground," he says, pointing to hollows and heaps in the foreground. "It suggested early bell pits, a very early form of mining going back to medieval times." Vertical shafts over coal seams, these pits assumed the shape of bells as miners expanded the base of the shaft once they hit the seam in the hope of getting both the coal and themselves out before the pit collapsed. Excavating some of the spoil thrown to the side, Williamson finds other clues: nodules of the underlying fireclay on which the coal vegetation grew; a few small cuboid shapes which he breaks open with his geologist's hammer. "Here we are," he says, rubbing the black exposed inside against the back of his hand. "It's dirty when you rub it on your hands. It's obviously coal."

Research at local libraries in Garstang and Lancaster confirmed the view from the field. There were numerous books mentioning coal mining in the Quernmore area in the seventeenth and eighteenth centuries. There was even a geological study of the area's coal seams by John Phillips, one of the founders of carboniferous geology. One-hundred-year-old, large-scale geological maps showed old coal workings in the area, and dropped other hints in such details as place names—a farm called Colliers' Gate, for instance. Finally, data from what was then the Coal Board confirmed Williamson's other findings: thin, poor-quality coal seams, what historically had been referred to as "boiler coal," had been worked in the area two to three hundred years before.

Coal usually meant methane—and there was one additional reason to believe it might be present and stored in natural underground reservoirs in the Abbeystead area. In the grit rock or sandstones of the area were thick layers of shale, some of which contained marine fossils. When decomposing, these fossils are known to form oil or methane which might have found a natural reservoir in the limestone. "Methane is light—it will migrate and, while the sandstone is dense, it is in fact very porous. It's an ideal reservoir rock," adds Williamson.

In court, Iain Williamson's testimony was given added credence by that of several other experts. But one witness had crucial practical evidence. Eddie Kluczynski had worked on the construction of the tunnel. But before that he had spent twenty-one years as a miner, including a spell actually testing for gas in coal mines. In fact, he had been mining coal ever since he came to Britain as a Polish refugee. He testified that during the boring of the tunnel he had reported the presence of methane three times. "I know it is gas, the smell of it and the terrible headaches I get. With twenty-one years' experience I can tell. I don't need equipment," he says. "I am sure something is not right there."

In the face of Kluczynski's repeated complaints, the constructors finally did some tests for methane. These, the HSE decided, were doubly flawed. First, the tests were done when the forced-draught ventilation was switched on in the tunnel. Second, the Draeger gas detectors used were crude instruments.

They could not differentiate gases, either natural gases or gases such as carbon monoxide being given off by the construction machinery. On three occasions the Draeger gas detectors did give positive readings yet each time no follow-up test that would have isolated methane was done. As the HSE concluded, the tests done "could not be taken to give a reliable indication of the presence or absence of methane."

Eddie Kluczynski was convinced that, as water engineers, not only did those in charge not know what they were looking for, but they were also looking in the wrong place. He was convinced that the gas was most prevalent where the tunnelers, who had worked simultaneously from both ends, made their breakthrough, about halfway along the tunnel. That was near where the build-up of methane that caused the explosion was believed to have come from. "He never finds this gas if he don't know where to look for it," Kluczynski insists. "I try to find it and prove there is gas there, but nobody wants to know."

But there was one other vital piece of evidence from Eddie Kluczynski. He had worked on a serious roof fall in the tunnel at about the midway point. Somewhere near here, perhaps about 250 feet down the tunnel from the Abbeystead Valve House, he recalls a big hole in the concrete skin of the tunnel through which water was pouring. It seemed to be an underground spring. Certainly the pressure was so great that the work gangs could not block it. He remembers accompanying a safety engineer or a charge man on a final inspection. "I ask him what about this hole and he says what about it. I say, well, there's water coming out and he says, it's tunnel water, we don't need to do nothing."

This may have been the major place of entry for the methane. Because it was near the roof fall and the breakthrough point, Eddie Kluczynski had observed the water emerging from this hole on numerous occasions, trying to work out where it was coming from, how they might block it. On many occasions the water ran "nicely, smoothly" as he puts it. But sometimes it was rather different. "From time to time, just bubble come, boom, boom, bubble pushing out. That tell me something is wrong here. Because water can't make a bubble like that on its own. There must be some kind of pressure there from somewhere."

Eddie Kluczynski had almost certainly seen the waterborne methane that would destroy the Abbeystead Valve House. It was another "unforeseeable" design disaster, this one dug as much as built, that was, in the end, ruled foreseeable; another act of God had been ruled an act of man.

OVERLOAD

NO TIME TO SCREAM

The Hyatt Regency

THE SWEEPING ARCHITECTURE AND GRAND DESIGN WERE THEIR KEY features. Fountained lobbies, huge atriums, glass elevators, balconies, and walkways, the Hyatt Hotel chain had always sought to stand out. "A place you talk about once you've been there," was the way one former president of the company had articulated Hyatt's marketing philosophy. In July 1980, the Hyatt Regency in Kansas City became one of the most spectacular additions to the company collection. Part of the Crown Center Complex, a major regeneration effort downtown, it was actually three buildings in one: a slim accommodation tower with 750 deluxe rooms and suites, a function block with the best restaurants, conference rooms, and ballrooms in the city, and, between the two, a huge fifteen-thousand-square-foot atrium.

Within a year the atrium, modeled on La Galleria in Milan, had become one of the social centers of Kansas City—a place to see and be seen. The steel and glass roof soared 50 feet above the floor and was straddled by three 120-foot-long suspended walkways or skywalks. These connected the two other build-

ings at the second and fourth levels on one side, and at the third
level on the other (see Figure 4). It was here that the Hyatt Regency's Friday evening tea dances took place, a hugely successful effort to recreate the Big Band atmosphere of a bygone
era. Attendances regularly topped fifteen hundred and it was
not just those who remembered the days of Duke Ellington who
turned out. Many of the regulars were young professionals and
Friday, July 17, 1981, was no exception. By 7 P.M. there was a
mixed crowd estimated at sixteen hundred in the atrium with
several hundred couples on the dance floor and scores more
watching from every vantage point, including the walkways.

Among the regulars were Walter and Shirley Trueblood.
Dancers since their teens, they had arrived, claimed their free
drinks, decided they were not up to the standard of those entering the competitive dancing and had taken a spectators'

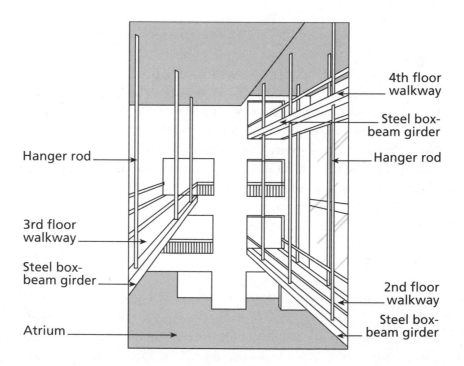

FIGURE 4
The atrium at the Hyatt Regency Hotel.

spot on the second-floor skywalk. "The band started to play 'Satin Doll,' which is still one of the best foxtrots ever written," recalls Walter. "It was an opportunity to watch good dancers. That is always great fun."

"As we were watching there was this loud crack, you could hear it over the sound of the band," says Shirley. "It felt as if the whole skywalk dropped, oh, six inches or so. We were standing at the railing along with some other people that we knew and Walt took my hand and said: 'You know, I really think we ought to get off this thing.'"

They never had time. For as James McMullin, a trial lawyer stepping out of the escalator onto the atrium floor at that moment, recalls: "There wasn't time to scream." Horrified, McMullin saw it all unfold in front of him. The first crack was followed by a series of sharp cracking noises, then a wrenching, screeching sound. The fourth-floor skywalk sagged, then split near the center. Unable to believe his eyes, McMullin saw both the second- and fourth-floor skywalks plummet downward, dozens of spectators falling with them. "Both walkways thundered onto the crowded floor—crumph, crumph. People and tables as far as a hundred feet away were mown down by flying glass and debris."

It all happened so fast, Shirley Trueblood did not even have time to realize the skywalk on which they had been standing had fallen. Her husband described it as like going down in an elevator in slow motion. Shirley's first memories are a series of sensual perceptions all at once: space, darkness, pain, and noise. "As soon as we landed we were trapped, pinned down. My legs were shattered from the knees to the hips and my face had been hit by some large chunks of concrete. It was pitch black. There were two people lying across my right arm. I'm sure they were dead. They never moved."

Walter found himself in a sort of tuck position on his left-hand side. But he too was trapped, unable to move anything but his left arm. "After the dust cleared, the first words I heard were from a man behind me who said: 'Let's pray for all those people that are trapped beneath us.' I confess I wasn't thinking about the people trapped beneath us. I was more concerned about where Shirley was and how I was going to get out of this. The next words I heard were Shirley's. I knew then that she was alive and at least we could converse lying under the debris."

Most of those directly underneath the skywalks were killed instantly. They ranged in age from a primary-school-aged girl to an eighty-year-old man. Eighteen married couples, most of them dancing where Walter and Shirley preferred to dance because there was more room under the skywalks, died together. But for many, their fate was decided by a change of movement or activity in the split second before disaster struck. Some walked into the disaster, some walked out. In the words of one young hotel employee who watched the skywalks land just seven feet from him: "It was unbelievable how, in that one terrible instant, so many lives were changed or ended."

Choking on plaster dust, James McMullin ran toward the fallen walkways. Cries and moans could be heard from beneath the huge mass of twisted steel and concrete. Along with others he tried to pry up the concrete. It was pointless. Each of the thirty-six-ton skywalks had broken into four. McMullin had served as a Marine officer in the Second World War but the carnage on the floor of one of the best hotels in Kansas City was worse than anything he had ever seen in battle. "Arms, legs, and mangled people protruded from under the wreckage. We'd pry and dig but we couldn't get to them. Men were crying in frustration, including me."

The first fire crew on the scene took a few moments to appreciate the quantity of heavy lifting equipment that would be required. Before long an emergency appeal went out. Workers from a nearby construction site were the first to respond, arriving with jackhammers, drills, crowbars, and oxyacetylene torches. As word spread, construction supply stores were opened and the firefighters and rescue teams invited to help themselves—arc lights, generators, compressors were all transported to the Hyatt. Eventually, the hotel's revolving doors would be pulled off to allow forklift trucks in. Cranes would eventually smash their seventy-foot booms through the lobby's glass exterior. The Hyatt Regency's lobby became a construction site and the real heavy lifting began.

The noise of the jackhammers terrified many of those stuck under the debris. "They sounded only inches over our heads and I was terrified they would suddenly plunge through and impale us," recalls Ed Bailey, a forty-three-year-old lawyer pinned under several layers of steel and concrete with his date, Shelley McQueeny. But the holes they made allowed the rescuers to push in air hoses and then themselves. Leading the paramedics, diagnosing, treating, and even amputating, was Dr. Joseph Waeckerle, head of emergency medicine at the city's Baptist Memorial Hospital. Leading the spiritual comfort, consoling, blessing, anointing, was Father James Flanagan, a fifty-seven-year-old Roman Catholic priest.

Walter and Shirley Trueblood had been entombed for more than three hours before rescuers reached them. "Someone

A computerized simulation of the scene on the dance floor at the Hyatt Regency Hotel as the flange of the steel box beams begins to crack.

touched my shoulder and said: 'Are you alive?'" recalls Walter. "That didn't register 'til years later when I realized this paramedic had been pulling out bodies." Numb with pain, Shirley was not sure she could hang on. "He was talking to us and I said I felt really faint. 'Do you think that's bad if I faint?' I asked. He said: 'Oh lady, please hang on, I'm almost there.'" With Shirley out, Walter was told he would have to be dragged out by his legs. The only problem was his broken pelvis. The pain was excruciating.

As the paramedics used Walter's belt and shoelaces to strap his thighs and ankles together, Father Flanagan appeared. "I'll never forget his ashen face as he leaned over me," Walter recalls. "I told him I was not of his faith and that I didn't think I needed the last rites but that the medics were going to have to tell me about that." At North Kansas City Hospital they soon did. It was a long list—a broken collarbone, a collapsed right lung, a broken pelvis, a shattered sacrum (the triangular bone at the back of the pelvis). From what the doctors could tell, Walter Trueblood would survive— but whether he would walk again and how well he might walk if he did was another matter. In fact, after surgery, both Walter and Shirley made full recoveries. Today it is only the cha-cha with its sharp sideways movements that gives the couple any problems on the dance floor.

It was 4:30 A.M. before the last living victim was pulled out from under the wreckage. But the rescue workers continued through the gray dawn, lifting layer after layer of concrete, hoping perhaps against hope for more miracles. At 7:30 A.M., after nearly twelve hours of work, the final section of the fallen walkways was hoisted skywards by the cranes. The rescuers had expected more bodies but were horrified by the number: thirty-one men, women, and children lay crushed to death. In their exhaustion, the emotional strain was too much for many. "The rescuers were visibly shaken. It seemed to drain the last ounces of their strength," recalls Father Flanagan.

The final toll was almost incomprehensible. Some 114 were dead and more than 200 injured, some of them crippled or maimed for life. The federal government soon labeled the disaster "the most devastating structural collapse to ever take place in the United States." In Kansas, it seemed even worse than that. More than 90 percent of the dead were from Kansas or Missouri. Everyone seemed to know someone who had been killed or injured or who had witnessed the event. As Kansas City's mayor, Richard L. Berkley, said: "It was like a death in the family."

After the Kemper Arena roof collapse into an unoccupied stadium two years earlier (see Chapter 1), it seemed as though Kansas City's luck with structural failure had run out with a vengeance. And at the local newspaper, the *Kansas City Star,* it was the Kemper Arena experience that was on the minds of journalists and editors facing one of the biggest stories of their lives. At Kemper there had been no published report into the causes of the disaster. Details had only emerged piecemeal over months and years as the whole thing had become bogged down in a legal morass of claim and counterclaim.

In the days following the Hyatt Regency disaster, a rerun of Kemper looked likely. The writs started to fly in the days immediately following the disaster, and victims, owners, builders, and designers all began to hire their own investigators in preparation for the long legal battle ahead. The editors at the *Kansas City Star* began to worry about how long it would take for the truth to emerge. Indeed, could any real truth at all emerge from such a process? Unwilling to be at the mercy of investigators working for all the different parties, the *Kansas City Star* decided to recruit their own engineering investigator and try and come to some of their own conclusions fast.

Wayne Lischka, a young structural engineer, was recommended to the paper by one of his colleagues. There were two problems in trying to involve Lischka in the newspaper's investigation: his status and access to the site of the disaster. One problem was solved over the following weekend. In response to a universal clamor, the media was informed it would be given limited access to the Hyatt Regency atrium on Monday. A plan was hatched to issue Lischka with a press

credential to accompany the *Star*'s leading reporters Rick Alm
and Thomas G. Watts. The only question now was: would
Lischka play ball?

On the Sunday evening Lischka was called by Darrell Liv-
ings, an assistant editor at the *Star.* Lischka had been highly
critical of the Kemper Arena investigation, feeling that no definitive conclusion had been reached. "So I felt I'd be somewhat
hypocritical if I didn't take part," he recalls, sitting in his sub-
urban office in Kansas City, a copy of the front-page article that
eventually emerged framed on the wall behind him. "I told him
I'd take part within certain parameters. I had to have editorial
control over everything and I had to be free to pull out if I
thought they were going in the wrong direction."

Reassurances given, Wayne Lischka met up with the jour-
nalistic team at the *Kansas City Star*'s offices and headed for
the Hyatt Regency. It was an emotional jolt: Lischka and his
wife had attended tea dances in the atrium. But from the moment the *Star* team arrived, it was clear that the tour was going
to be a big disappointment. The journalists were allowed to
view the atrium only from the second floor of the restaurant
area. They were a hundred feet or more from the scene of the
collapse and the wreckage, which was now all laid out on the
atrium floor.

The photographer's telephoto lens was the only option for a
closer view. Lischka pointed out the fractured joints and steel
tension rods he wanted photographed in as much detail as possible. The third-floor walkway was still in position, suspended
from the ceiling. With fixed bearings attaching it to the wall at
either end, and three sets of hanger rods suspending it from the
ceiling to connections in the steel box-beam girders that
formed the side of each walkway, it demonstrated the structural design of the walkways that had collapsed.

If the third-floor walkway illustrated the design, a brief look
through the photographer's telephoto lens showed the damage
done where the connections had been placed in the box beams.
The hanger rods from which the collapsed walkways had been
suspended from the ceiling were still in place—it was only the

Right: The bent hangar rod of the fourth-floor skywalk, its nut still in place. It had pulled through the flange of the steel box beams to which it was attached.

last six or eight inches that were raw and exposed. Some were bent, indicating the first failure points (see photo, right). With the nuts still clearly in place on the rods and no evidence of what structural engineers call "necking"—stretching so it becomes smaller in diameter to form a neck—it was clear that it was the connections rather than the rods themselves that had failed. For Wayne Lischka and the *Kansas City Star's* investigative team, the key question was why.

THERE WERE ALREADY PLENTY OF THEORIES BUZZING AROUND KANSAS City. Roger McCarthy, heading a four-man team from Failure Analysis Associates (FaAA)—leaders in the field of engineering failure investigation—had already heard three prominent theories by the time Wayne Lischka and his colleagues got access to the scene. One: the fourth-floor walkway was simply overloaded. Two: the materials used in the walkways were inferior. Three: this was the ultimate case of a collapse due to resonance—the people on the walkways had been stomping in time with the music. Continued resonance, as all structural engineers know, can destroy the soundest of structures.

"We performed some rough calculations on what the natural frequencies of the walkways were," says Roger McCarthy. "They were much higher, more cycles per second than probably your best go-go dancer could generate. So there wasn't much chance that those kinds of generation were going to be produced at a tea dance in Kansas City." The resonance theory was rejected out of hand. However, McCarthy and his team did believe that the walkways were overloaded. But they were overloaded only relative to their design. They were, as structural engineers say, underdesigned. "On a properly designed walkway, people standing back to front could not have overloaded. It wasn't even as if the people on the walkway were the most significant factor in its overall weight. They were not," asserts McCarthy. "It was simply that the basic design of the walkway was so close to its safety limits that those limits were breached by the addition of relatively few people with the result that the connection pulled through and failed."

It was this underdesign that Lischka now set out to analyze. To do so he needed to see the plans filed at City Hall when the Hyatt Regency was built. But City Hall had declared the plans off limits—until the following day, Tuesday. The *Kansas City Star,* the only news organization with a full-time employee at City Hall, managed to get preferential access on Monday night.

Wayne Lischka's was one of four names placed on a list deposited with the security officers at the entrance to City Hall that evening. With three journalists from the *Star* he was shown to a mundane office room where a sheaf of plans was spread out on a table. In the end it took Lischka longer to find the relevant sheets—the sheets that dealt with the walkways' connections—than it did to spot the key to the whole disaster. "It was obvious immediately," Lischka says of his eureka moment.

The original plans filed at City Hall showed the second- and fourth-floor walkways each hanging from six one-and-one-quarter-inch steel rods hung from steel trusses in the atrium's roof. They showed a pair of holes drilled through the flange or projecting end of each box beam with the steel hanger rod threaded through to suspend both the second- and fourth-floor walkways. In this design the load of each walkway was supported by nuts and washers screwed onto the steel rod at the level of each walkway even though both walkways were hanging on one rod (see Figure 5, top left).

That was not what Wayne Lischka thought he had seen at the Hyatt, something the developed photos of the box beams taken on the telephoto lens would later confirm. Lischka believed he had seen a design based on two rods. Each end of the

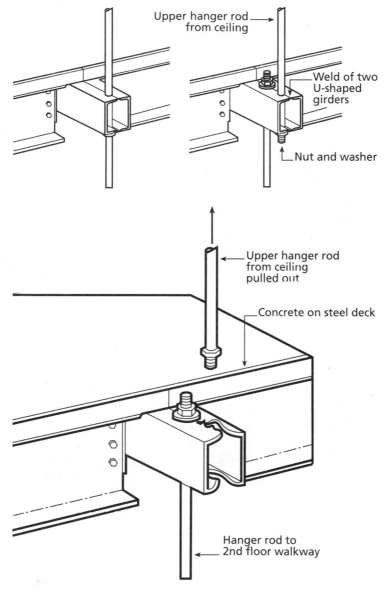

FIGURE 5

Top left: The original design of the fourth floor walkway connections.
Top right: The connections as they were built after the design was
adapted. *Bottom:* The adapted design fails as the strain becomes too
great and the upper hanger rod is pulled through the top and bottom
flange of the box beam.

fourth-floor box beams that extended out of the walkways had two pairs of holes through both sides of the hollow metal box beams: one at two inches from the end and another pair at six inches from the end. The two upper hanger rods, those suspending the fourth-floor walkway, went through the outer holes at both ends, supported only by nuts and washers underneath. Two separate, lower, hanger rods went through the inner holes of each fourth-floor box beam and suspended the second-floor walkways, again with nuts and washers top and bottom (see Figure 5, top right).

The design had been changed and it was pretty obvious why. The first design as submitted on the plans was simply impossible to construct. "You would have had to take a seventeen thousand-pound walkway and basically slide it up twenty feet of threaded rod fixed to the ceiling without damaging the threads, then tighten the nuts. It was just not feasible," observes Lischka. "The original design wasn't practical and was adapted to a double rod. There would have been nothing wrong with that if they had put a plate underneath to distribute the load instead of using just washers."

The effect of the design change was to double the load on the connections on the fourth-floor walkway. With both walkways carrying the tea-dance spectators, the fourth-floor walkway was carrying not only its own live load but that of the second-floor walkway as well. The obvious analogy was two people hanging on a rope. If both are hanging onto the rope separately, they have only to support their own weight. If, however, one person is hanging onto the legs of the other who is hanging on the rope, that person is supporting his own and the other's body weight. Their connection to the rope—the top person's hand grip—will come under much greater strain.

Realizing that the alteration in the design of the suspending connections was probably the key to the whole disaster, Lischka asked the *Star*'s office to dispatch a photographer. Armed with photographs of the plans, the team now returned to the newspaper office to put together a front-page scoop. Here Lischka was able to make comparisons between what he had seen on the plans and the blown-up photographs of the joints, connections, rods, and pieces of walkway that he had seen earlier that day at the site of the disaster.

While studying the photographs of the debris, Lischka noticed something. The holes for the hanger rods ran through the middle of the box beams, just where the two side-on U-shaped girders that formed the box beams were welded together to form the actual box. The holes had thus been drilled through the weakest point of the box beams. The photographs showed how the metal here was bent upwards as if it had been under strain for some time. Eventually it became sufficiently weakened to create a large enough hole for the lower nut to pull through the bottom flange of the box beams and stop momentarily at the upper flange (see Figure 5, bottom). This was probably the initial six-inch downward jolt that Shirley Trueblood had felt on the second-level walkway moments before the final collapse. "The walkways were starting to collapse from the day they were constructed. It was failing from day one," concludes Lischka.

Sustained by adrenaline and the excitement of knowing they had cracked the story, the *Kansas City Star*'s team worked all night. Rewrite followed rewrite as sub-editors, editors, and the art department all had their say. Eventually everyone was happy. "We finished writing about 6 A.M. on the Tuesday morning and the paper hit the streets that afternoon," recalls Lischka. The front-page scoop of July 21, 1981, headlined "Critical design change is linked to collapse of Hyatt's skywalks," complete with photographs and artists' illustrations, caused a sensation. Lischka knew nothing about the reaction to his work. When the paper's scoop hit the streets, he was fast asleep in bed.

Within days, City Hall released the plans that Lischka had inspected. The third-floor walkway, showing the deformation in the metal box beams that the young structural engineer had identified in the collapsed walkways, was removed from the hotel atrium. Structural engineers had confirmed Lischka's view—it was almost as dangerous as the two walkways that had collapsed.

In the succeeding months those investigators who did see the debris close up, and were able to carry out tests and experiments, did little more than amplify Wayne Lischka's initial findings. McCarthy and his team were soon joined by investigators from the National Bureau of Standards (NBS), asked by Kansas City officials to establish the technical causes of the collapse. Several things were immediately obvious to all the investigators now on the job.

Even the initial design—with continuous rods suspending both the second- and fourth-floor walkways—did not meet Kansas City Building Code specifications. The NBS report concluded that the ultimate load capacity of the original connection detail would have been "approximately 60 percent of that expected of a connection designed in accordance with the Kansas City Code." However, although very close to the failure

Rescue workers begin searching through the rubble of the collapsed skywalks at the Hyatt Regency Hotel in Kansas City, the worst single construction disaster in the United States.

margin and certainly unsafe, the connections, as originally designed, may well have held on the night of the disaster.

As redesigned, without any alteration to the box beam, connection, or hanger rod, there was no hope. Even the dead load of the walkways—their own weight—was more than that specified in the design, the NBS decided. By adding up the weight of all the components of the walkways as specified in the plans, then adding up the actual weight of the components recovered from the collapse, NBS investigators Ed Pfrang and Richard Marshall discovered that the actual dead load was thirteen hundred pounds, or about 8 percent more than the computed dead load. A corrugated steel deck and a concrete topping not shown on the original drawings made the difference.

The extra weight took the walkways even closer to the edge of their safety limits. They had only minimal capacity to resist the dead load and virtually no live load capacity. This could be demonstrated by the actual live load on the night of the collapse. By chance, the local television station KMBC-TV had been filming the tea dance. Although the crew were changing batteries at the actual moment of the disaster, shots of the second- and fourth-floor walkways taken moments before allowed investigators to conclude that there were about sixty-three people on them at the time of the collapse. That gave an approximate live load of 9450 pounds. The live load required by the Kansas City Building Code was 72,000 pounds per walkway, in other words a total of 144,000 pounds. The walkways had collapsed with a live load of about one-fifteenth the load they should have been able to carry.

None of the investigators believed that both sets of connections were anything other than the most elementary mistakes. "Almost any practicing engineer could have seen that the skybridges would not meet the Kansas City Building Code," concludes Ed Pfrang of the NBS. It was simple to solve because it was so simply wrong, Wayne Lischka maintains. "This was a very simple failure. There was no sophistication here. It didn't require a brilliant engineering mind. It was just obvious from day one."

A check of the records revealed something of the system that had allowed such a basic design flaw to slip through the safety net procedures not once but twice. The failure of the skywalks was ultimately a failure of the people involved in the design, construction, and inspection system who had allowed it to happen. Further investigation, and endless legal action, would eventually reveal three things. One: the Hyatt Regency was a troubled project that had experienced serious failures while in construction. Two: chaotic lines of communication and chains of responsibility between designers, architects, and constructors had been the norm. Three: woefully inadequate inspection had failed to catch up with any of this.

The structural drawings for the Hyatt Regency had been prepared by the project's appointed consulting engineers, GCE International Inc., in 1977. These drawings, including the continuous-rod connection plans for the walkways on those numbered S303, S304, and S305, were personally sealed by one of GCE's partners, Jack D. Gillum. As was the practice, the structural drawings were sent to the fabricator, Havens Steel Company, who had been subcontracted by the general construction contractor, Eldridge & Son Construction Company, to fabricate and erect the atrium steel, in order to draw up what were known in the trade as the shop drawings. The specifications were to be based on the American Institute of Steel Construction (AISC) standards used by fabricators.

In December 1978 or January 1979, an unnamed detailer at Havens Steel Company rang Daniel M. Duncan, the structural engineer handling the project at GCE, and told him that he did not think the second- and fourth-floor walkways could be suspended from a single continuous rod. He preferred a two-rod system. Duncan admits telling the detailer he had no problem with that. There is no evidence that any calculations to check the changed connection for loading and structural integrity were ever done.

Some forty-two shop drawings, including Shop Drawing 30 and Erection Drawing E3 which detailed the two rod connections, were sent back to GCE on February 16, 1979. During a standard shop drawing review at GCE, a technician noted the discrepancy between the structural drawings and the shop drawings and raised the issue of the changed connection and the strength of the rod with Daniel Duncan. Duncan stated that the two-rod system was basically the same as the single rod. He says he then looked at the shop drawing but did not "review" the box-beam connection. Ten days after their arrival at GCE on February 26, the relevant shop drawings were stamped with the company's review stamp: "Reviewed only for conformance with the design concept and compliance with the information contained in the contract documents."

The structural drawings were those reviewed by Wayne Lischka and his *Kansas City Star* investigative team at City Hall just three days after the disaster. Some time later the *Star* managed to obtain the shop drawings. They had all been stamped and initialed by the hotel's architect, PBNDML Architects Planners Inc., and the general contractor, Eldridge & Son Construction Company. In other words, all three major parties to the design and construction of the walkways were aware of the change before building actually got under way.

NO ONE, NOT EVEN THOSE WHOSE JOB WAS INSPECTION, PICKED IT UP. IT would be logical to assume that if it was missed on paper, it would at least be picked up in practice. Not by Kansas City building inspectors, it seems. Analysis of logs, time sheets, and inspection paperwork by the *Kansas City Star*'s investigative team revealed one of the reasons why. The city building inspectors spent only eighteen hours and twenty-one minutes inspecting the hotel's construction during more than two years of work on the site between spring 1978 and summer 1980. Officials in other cities assured reporters that their inspectors would have spent much more time on a project of similar magnitude and complexity. The inspection system was as "underdesigned" as the atrium walkways, it seemed.

As if this was not enough, the structural engineers claimed to have done a special inspection of every detail in the atrium. On October 14, 1979, a somewhat difficult construction project seemed to touch rock bottom when twenty-seven hundred square feet of the atrium roof collapsed when still under construction. The incident was serious enough for the owner to hire his own independent engineering firm to investigate the cause. Jack D. Gillum assured the owner he would check every detail in the atrium. A report was sent to the architects in November and, at a meeting in December, GCE claimed they had reviewed every connection in the atrium. In 1984, a judge decided that the engineers had never checked the fourth-floor walkway connection despite not only claiming that they had but requesting an additional fee for doing so.

All this emerged in a landmark negligence suit in 1984–85 between the State of Missouri and GCE International Inc. In a 442-page report, Judge James B. Deutsch found Daniel M. Duncan, Jack D. Gillum, and GCE International Inc. guilty of gross negligence, misconduct, and unprofessional conduct in the practice of engineering. GCE and its structural engineers had claimed that the steel fabricator was responsible for connection design. The State had argued that "total responsibility for structural safety rests with the structural engineer of record no matter what...." The State revoked the engineering licenses of those concerned and they were subsequently suspended from membership of the American Society of Civil Engineers (ASCE) for violating the organization's code of ethics.

FOR YEARS, STRUCTURAL ENGINEERS IN THE UNITED STATES AGONIZED over the lessons of the Hyatt Regency collapse. The need to establish much clearer procedures and responsibilities was obvious; the absence of effective project peer review and what engineers termed "constructability review" was perhaps fatal. But in the end it was hard to avoid the conclusion that negligence was to blame. In a compelling paper published in *The Construction Lawyer* in August 1986, Robert A. Rubin and Lisa A. Banick, both members of the ASCE themselves, argued that Duncan and Gillum were "not hapless victims of a failed system." They pointed out that the right people did ask the right questions at critical times. There were several opportunities to correct the error. The design and change of design did not, in their words, "fall through the cracks."

Rubin and Banick argued that the structural engineers' failure to discover the problem was "inexcusable by any standard." They also argued that an attitude of cavalier negligence was to blame. This in turn they blamed on complacency. "Some succumb, some do not; most are just plain lucky in that they do not get caught.... Complacency is a human failure," the authors argued. "It creeps into a professional's approach to practice as the newness, excitement, and other early rewards of the profession fade. The professional becomes indifferent and stops worrying and agonizing. He takes shortcuts and gets away with it and then takes more shortcuts. It becomes a way of life."

Guarding against this complacency in structural engineering, both personally and professionally, was perhaps the biggest single challenge to preserving public safety in buildings. To date, fortunately, there has been no worse failure to keep it at bay than that at the Hyatt Regency Hotel in Kansas City.

5

CROOKED CONSTRUCTION

Sampoong Superstore

T HE MID-1990S WERE A BOOM TIME IN SOUTHEAST ASIA AND, OF ALL the region's tigers, none had run harder, grown faster, become stronger than South Korea. Twenty-five years of annual average growth rates of more than 10 percent had transformed a developing nation into a developed one within a generation. And nowhere demonstrated that transformation more starkly than downtown Seoul, the capital. Six-lane highways, high-rise office blocks, plush department stores, and fashionable restaurants had mushroomed amid the billboards and neon lights that, by 1995, swathed the city.

No two sectors of the economy grew faster than construction and retail—buildings to accommodate the companies and people forging the economic transformation, and up-market stores to cater for their developing tastes. The Sampoong department store was part of both. The nine-story building, five upper and four basement floors, had opened in the up-market Kangnam area of the city in December 1989. It offered an unrivaled range of clothing, furniture, crockery, and other luxury goods and did well. By the mid-1990s the store was taking in the equivalent of more than half a million U.S. dollars a day.

The pink-fronted structure became something of a landmark, as did the building's distinctive lay-out—two wings, north and south, connected by an atrium lobby. But the gloss and glitz of the neon-lit chrome-and-marble façades of the interior hid a checkered history. The store had been built on a landfill site, and Woosung Construction, one of the country's largest construction firms, did the foundation and basement work before Sampoong's in-house contractors, Sampoong Construction, erected the superstructure.

Woosung had apparently balked at making significant changes to the building plans, including the addition of a fifth floor. But in the hands of Sampoong's own construction company, the design changed even more radically. What had been planned as an office block became an open-plan department store. The upper floor, which had originally been earmarked as a roller-skating rink, became a traditional Korean restaurant. The change of use meant much stricter safety guarantees as regards fire, air-conditioning, and evacuation. As a result, the final structure apparently met all the building code provisions for its new use as a department store—yet the revised design bore hardly any resemblance to that originally drawn up.

For five and a half years business thrived. In June 1995 the store passed a regular safety inspection. But within days there were signs something was seriously wrong: cracks spidering up the walls in the restaurant area; water pouring through crevices in the ceiling. On June 29 structural engineers were called in to examine the building. They declared it unsafe. Company executives who met that afternoon decided otherwise. They ordered the cracks on the fifth floor to be filled and instructed employees to move merchandise to a basement storage area.

Word quickly spread among the staff that something was up. "When I came back from lunch I heard that the roof on the fifth floor had collapsed and that people had been evacuated," recalls Yoo Ji-Hwan, a nineteen-year-old woman working in the

The Sampoong Superstore in Seoul. A "change of use" from office space to shopping complex led to unsafe alterations and one of the most deadly collapses the world has ever seen.

crystal department on the second floor. "I asked if all the other floors were OK and just thought it would pass. No one thought it was a big deal." On the second floor, where she sold children's wear, Park Seung-Hyun heard something similar. "I heard rumors that the restaurant floor, the fifth floor, had sunk and that they couldn't turn the air-conditioning on because the vibrations would run right through the building. I didn't really pay much attention to the rumors."

It was just before 6 P.M. on June 29 when disaster struck. The store was fairly busy, but the customers were concentrated in one or two departments—housewives shopping for the evening meal in the basement, families and business people eating in the restaurant on the fifth floor. "I was chatting to colleagues when we felt vibrations," recalls Yoo Ji-Hwan. "As they became more frequent our boss in the crystal section

said we should evacuate quickly but she just stood there. I
started to run to the corner but had only taken a few steps before I fell." It all happened so quickly that the first thing Park
Seung-Hyun remembers is people running all over the place
screaming. "I just started running with them without thinking
but I didn't get very far before I was hit on the head by a concrete block from the ceiling. I passed out."

The building was collapsing in the most spectacular style.
"It just folded as if it was being destroyed by a demolition
crew, the way you see on television," said Park Min-Soo, a cab-
driver waiting at a traffic light in front of the store. "It just went
in a matter of seconds—ten, maybe." The five stories of the
north wing, some three hundred feet in length, had just subsided into the basement, with the centers of the floors going
first, the side walls and corners folding in on top. The only
thing left standing was the building's glitzy façade, still
proudly boasting the name of a store that no longer existed.

As the walking wounded ran from the building through the
dust and smoke into the street, the emergency services began
to arrive amid a chorus of sirens and shouts. It was soon clear
that there was little hope for anyone in the middle of the floors
that had collapsed; rescue efforts quickly focused on the sides
of the building and the basement. Rescuers could often hear
the cries of those trapped under the rubble but the scale and
complication of the task threatened to overwhelm them. Fires
and toxic smoke impeded access; the thousands of tons of concrete and steel debris were still dangerous. Even though rescue
attempts were quickly reinforced by heavy lifting equipment,
sonic listening devices, and by U.S. military personnel, urgent
appeals were soon being broadcast for basic rescue equipment—jackhammers, steel-cutters, and arc lights.

Park Seung-Hyun came round to find herself enveloped in
darkness and silence. It took a little time to realize that the
building had collapsed and that she was trapped, somewhere
in the basement under hundreds of tons of rubble. Gradually
she became aware of someone else. "I heard a familiar voice
through the concrete, someone I knew from the children's wear

department. 'Help me, help me,' she cried." As time went on, the young assistant became aware of other voices, other friends. "I called one of their names, Hye-Jung. She recognized my voice and called out my name."

Park Seung-Hyun started to feel her way around, moving bits of debris, looking for a chink of light that would point the way out. There was none and the heat was stifling. Exhausted, she fell asleep. On waking she could hear Hye-Jung sobbing. "She kept asking for help, saying she was badly hurt and in pain. I asked her where the pain was and she said she thought something had gone through her side. I told her to stay calm, not to scream. But she was in so much pain she just couldn't cope."

Yoo Ji-Hwan also survived the collapse. "I opened my eyes but couldn't see anything. At first I thought my eyes would adjust, but they didn't. It was total darkness." She was completely unaware of the enormity of what had happened, surmising that a wall had probably fallen down and she would be rescued fairly quickly. "Most people have never heard of a building collapsing. I just couldn't imagine it. I was thinking what I'd do once I was rescued. Go home, get a shower, probably have the next day off as they did the repairs. I just thought it had been my unlucky day."

As time passed, Yoo Ji-Hwan began to realize she might have to help herself. She began to feel her way around. Concrete debris, then shelves and, in a corner, knives, scissors, broken glass—the contents of the household utilities department that had plummeted down toward the basement with her. Then she found a lighter. "I flicked it on but there was nobody there. I was alone. I saw I was injured in a few places and began to feel all over my body. I felt something wet and realized it was blood." She worked out that she was bleeding from the face and from what seemed a sizeable wound in her back. Now terrified that she might bleed to death before being rescued, she took off some clothing, found the scissors and improvised a bandage.

The voices and noises of the rescuers came and went. Yoo Ji-Hwan and Park Seung-Hyun tried to attract their attention by

banging with bits of concrete debris and even a piece of piping, but no one could hear them. Both young women drifted in
and out of sleep. "People ask how can you sleep in such a condition but you can actually sleep better in a situation like that.
By sleeping you escape, you are able to deny the reality," explains Yoo Ji-Hwan. "I slept, woke up, slept, woke up tens of
times," recalls Park Seung-Hyun. "I dreamt, too. A Buddhist
monk I was close to appeared in one of my dreams and showed
me a painting of an apple on rice paper. I thought it was a good
omen. In another dream I saw my cousin at a swimming pool. I
asked her what the date was. She told me the date...I thought
five days had passed since the collapse."

Outside, the death toll was mounting as day after day, night
after night, the layers of concrete and steel were peeled back.
But rescue teams continued to pull out the living as well. During the night of June 30, nine people were rescued alive. More
were extracted the following day. But the dangers from collapsing debris, fires, and toxic smoke—dangers to both the victims and the rescuers—were equally clear. Two of those
rescued on July 1 died on the way to hospital. "I just discovered five people dead in one place," one rescuer told the press
that day. "They were alive when I was there about four hours
ago. They were suffocated by toxic smoke."

Later that same day rescue workers made their biggest breakthrough to date when they secured access to a basement stair-
way leading down to a third basement-floor locker room.
Hearing shouts through the slabs of concrete blocking their path,
they spent ten hours cutting through the debris. Some twenty-
four cleaning staff were still alive in the basement room. Many
were older people; some severely dehydrated. In the end all
twenty-four managed to slip through the biggest hole rescuers
could cut, just twenty inches in width, by greasing themselves
with cooking oil. "We often held hands and just persuaded each
other not to lose hope of being rescued," Han Kyung-Sok, a fifty-
three-year-old maintenance man, said of their ordeal.

Following the rescue of the cleaners on July 1, it seemed as
though the search would be abandoned. Safety officials de-

cided the ruin was too dangerous. In particular, an elevator shaft that had been left standing after the collapse was tilting perilously. As more than five hundred relatives waiting at the site learned of the news, pent-up grief turned to anger. Scuffles broke out with riot police as tearful relatives pleaded with officials to resume the search. On Sunday morning, with new shoring at the base of the elevator shaft, the relatives prevailed. Within hours another survivor had been pulled from the rubble. That afternoon, twenty-two-year-old Lee Eun-Young was extracted from a small air pocket, bleeding and exhausted.

When another store employee, Choi Myong-Suk, twenty-one, was rescued after nine days, he seemed certain to be the last survivor. The weather had changed while he was trapped and he had survived by drinking rainwater. Inexorably, rescue became recovery, with the death toll mounting ever higher as more bodies were extracted. Yet, although lapsing in and out of consciousness, both Park Seung-Hyun and her colleague Yoo Ji-Hwan were still alive. And as the debris was cleared and rescue workers moved, unknowingly, closer, both shop assistants faced a new threat—the collapse of their own small survival spaces.

"The ceiling was gradually sinking and eventually closed off all the space to one side of me," recalls Yoo Ji-Hwan. By then, she could no longer stretch out. "At one point the ceiling literally came this close to my nose," she says, placing her hand just inches from her face. It was all the more frustrating because after some time she started to hear the voices of the rescue workers again. "I would call out, bang anything I could get my hands on. At times you could hear everything...it felt like you could reach out and touch them, it seemed so close. But there was no response. You start to lose hope that you will be rescued. Your thoughts just oscillate violently...I won't die.... Could I die?...Yes...I probably will die."

The disturbance of the debris at her feet was in fact rescuers and it was here, on the thirteenth day, that someone finally appeared. "Suddenly there was a hole above my feet and I heard a voice. 'Is there anyone down there, is anybody alive?' I answered but they didn't seem to hear me. I didn't think I was

A total of 499 people died under the rubble of the north wing of the Sampoong Superstore.

that weak but they couldn't hear my voice," Yoo Ji-Hwan recalls. "They asked me to move my feet so I moved my feet. I had to let them know I was alive. What if they thought I was dead and passed by? I moved my feet a lot."

As Yoo Ji-Hwan was extricated into the soft drizzle and cloud of an overcast day and rushed to the hospital, Park Seung-Hyun remained in her shrinking underground tomb in the second basement level on the northern edge of the collapse site. "The ceiling kept coming down, the space around me kept getting smaller," she recalls. "I was on my stomach for a long time. I wanted to turn around but couldn't because there just wasn't enough space."

It would be four more days before, on hearing voices again, she managed to pick up a metal pipe and bang on the concrete screaming for help. All she heard was the echo of her own voice, until a forklift truck seemed to start working right above her. Although there was even less chance of being heard over the roar of the engine as it shuttled back and forth shifting debris, she tried again. Suddenly the engine stopped. The operator had heard a faint moaning sound. "I heard someone say: 'There is someone down there!' and I heard digging right above where I was," recalls Park Seung-Hyun. "I saw his face and my only thought was I was going to live. I could finally leave this darkness." Park Seung-Hyun had been buried for seventeen days.

BOTH PARK SEUNG-HYUN AND YOO JI-HWAN, THE LAST SURVIVORS TO BE rescued, were to make full recoveries but by the time they were hospitalized public outrage had reached fever pitch. With the final death toll put at 498, everything was pointing to poor, possibly criminal, design and construction work. Police had already questioned Sampoong's chairman and president Joon Lee and his son, Han-Sang Lee, the company's chief executive. A criminal investigation was now under way. It would need as much evidence as possible from the structural engineers now poring over the ruins and the plans.

Leading the way were Lan Chung, professor of civil and structural engineering at Dan Kook University in Seoul, and Professor Oan Chul Choi, head of the department of architecture at Soongsil University. Both professors began looking at the debris while the rescue efforts were continuing, and photographs were one of their first priorities. The first thing that was immediately obvious was that the Sampoong department store had been a flat-slab structure, without cross beams or a steel framework. "In such a building the fitting of the floors and ceilings is absolutely crucial—they are the framework, so it has to be erected exactly as designed. It has to be perfect," says Professor Choi. "Without cross beams you are effectively missing one form of load transmission," says Professor Chung. "The weight of the ceiling and the floors above is transmitted directly to the columns between floor and ceiling—an important redundancy factor is missing."

The flat-slab design explained why the building had collapsed so completely, like a house of cards. The domino effect was common in such structures, as the collapse of Ronan Point, the apartment tower block in London, proved back in 1968 (see Chapter 7). But that did not explain what had actually caused the collapse. Had there actually been some fundamental fault or alteration to what was after all a notoriously sensitive design system?

There were repeated rumors that the concrete used in the building had been substandard. Indeed, a section of badly honeycombed—holed and pitted—concrete was visible along the lone wing wall of the building still standing. The investigators took dozens of concrete core samples from the site and tested their tensile and compressive strength in the laboratory. The concrete was not the best but it was by no means the worst either, and ultimately, investigators decided, it was impossible to determine how much it may have been weakened during the actual collapse itself.

The weakness of the concrete may have compounded the disaster and hastened the collapse but it was certainly not the cause. The same seemed true of another possible cause—the ground structure. Although the Sampoong department store had been built on a landfill site, the foundations and basement structures built by Woosung Construction seemed to be sound. They had survived the collapse well. Bore samples revealed that the building had been founded on rock.

A design review of the plans was the investigator's next step. The consequences of the conversion from an office block were immediately obvious. The owners had needed to insert escalators on every floor. Punching holes in the slabs making up each floor and cutting out some of the support columns had certainly weakened the structure.

Similarly, to conform with fire regulations the owners had been obliged to fix shutters, some of which had been attached to the wall columns further reducing their load-bearing capacity. But the floor columns had also been weakened to conform with fire regulations. In fixing fire doors to allow each department to be isolated from its neighbor, huge chunks of the columns had been cut away to make them fit. The problem had been as simple as incompatible shapes: the steel doors were rectangular, the columns round.

But it was on the fifth floor—the additional floor that Woosung had refused to build—that the crucial changes had been made. The conversion to a restaurant had meant a huge increase in weight. Because customers sit on the floor in traditional Korean restaurants, a three-foot-deep underheated concrete floor had been added. That meant an additional layer of cement, over a foot thick, across the floor. Refrigerators for the restaurant also increased the load.

Furthermore, the whole design of the fifth floor was incompatible with the lower stories. There were more windows, which weakened the walls relative to those on lower floors. There were also more support columns, whose design and placement were irregular. Because they were not positioned above the columns on the lower floors, the increased load was being irregularly transmitted to the lower floors. The ceiling of the fourth floor, rather than the columns, took the strain.

This was to prove crucial because on what was now the roof of the building another critical weight feature had been added. Three huge water-cooling blocks, the basis of the whole department store's air-conditioning system, had been built— again as a result of the building's change of use—in August 1993. In winter, they were empty and weighed fifteen tons each, but in summer, when full of water, they were twice as heavy. It was actually no accident that the load had finally proved to be too much in midsummer when the cooling blocks were working flat out, Professor Chung decided.

"The cooling blocks were actually designed to be on the ground and would not have been in use until the summer," he explains. When they were placed up there they should have calculated the load-bearing capacity of the columns to see if they could take it, then added new columns as necessary. Of course they didn't. They just added some roof slabs, about ten inches thick, which further increased the load."

Professor Chung and Professor Choi did some detailed calculations and comparisons with the section of the store which
had not collapsed and which incidentally had no air-
conditioning blocks on it. "All central heating systems like this
require much larger design-load specifications," explains Professor Chung. "We worked out that with cooling blocks like this
on the roof it would have required a sustained design-load capacity of about four hundred kilograms per square meter. This
building was designed with an ordinary roof load-bearing capacity of one quarter of that."

The air-conditioning cooling blocks had apparently been
placed on the roof because the noise they made was thought
likely to draw complaints from residents of the Kangnam suburb. In fact, people living in nearby apartments had complained even when the blocks were on the roof. The owners'
response only compounded the weakness problem. They decided to reposition the blocks, moving them from the back of
the building to the front. "Instead of using a crane to completely lift them off the ground they simply slid the cooling
blocks across the roof on some slabs," says Professor Chung.
"This weakened the whole of the roof and cracks were clearly
visible afterward. We concluded that a lot of the structural
damage was done when they moved the cooling blocks."

Interviews with staff confirmed to investigators that cracks
of up to an inch in width had first appeared beneath the air-
conditioning blocks just where the columns met the roof
slabs. It was what structural engineers term punching shear-
failure—the column punching through the roof slab as it became unable to take the load. Adjacent columns were then
unable to take the additional load and progressive collapse
followed as one after another punched through the ceiling. As
the roof sections with the water-cooling blocks failed, successive floors and ceilings below were unable to take the weight
and followed suit.

A whole series of design deficiencies now came into play
even on the first four floors, the regular part of the construction.
Some columns specified in the designs as being thirty-one to

thirty-five inches thick were in fact less than twenty-four inches; some contained eight steel reinforcing rods rather than the sixteen specified. Ceiling-slab dead loads were hugely undercalculated, investigators decided, some being based on four-inch slabs when some were three and even four times thicker. Reinforcement was often missing, and the connections between floor/ceiling and outer walls were, as so often in flat-slab system-built structures, inadequate, with tie-ins and over-laps failing to meet design specifications. The flat-slab construction span between each column, nearly thirty-six feet, was just too large, investigators decided—the result of trying to maximize sales space.

As a result, in their final report Professor Chung and his colleagues blamed the Sampoong department store collapse unequivocally on "human ignorance, negligence, and greed." The prime cause, they said, was the "illegal alteration of the architectural design and usage purpose of the building." They cited the negligence of supervision by the planning authorities and the refusal to act on any of the indications of structural problems by the management as crucial contributing factors to the disaster. Cracks and leaks had been appearing in the building for more than five years, they stated. The addition of the air-conditioning cooling blocks had simply added to the problem.

The findings caused a public outcry. The revelations that the store's senior executives had ignored structural experts and refused to evacuate the building on the day of the collapse—despite leaving the store themselves—was particularly galling. But the system that allowed the tragedy to occur seemed to be part of a pattern of deep-seated corruption at all levels of the construction business. The disaster was new only in scale. In October 1994 a section of Seoul's Songsu Bridge over the Han River had collapsed during the morning rush hour. Thirty-two had died. In April the following year, 101 people had died in a gas explosion at a subway construction site in the southern city of Taegu.

In the wake of the Sampoong disaster, Koreans began to ask themselves what price they were paying in their headlong rush for economic development. Commentators pointed out that all these disasters had followed the building boom launched in the mid-1980s, a boom propelled not just by economic growth but by the staging of the Olympics in Seoul in 1988. With foreign firms excluded from tendering for construction contracts, supervision had become lax, corruption rife, Korean newspapers claimed. "With too few construction firms trying to cope with increased demand, corners were cut and safety regulations ignored," claimed *AsiaWeek* in July 1995.

That assertion was borne out by a government survey of the state of high-rise construction in South Korea. It concluded that 14 percent of all high-rise structures in the country were unsafe and needed rebuilding, 84 percent needed repair work, and only 2 percent of such buildings met government standards. "As a Korean structural engineer I feel embarrassed," concludes Professor Choi.

In the case of the Sampoong Department Store, at least, action was taken, action that shed some light on the system that had allowed it all to happen. After months of demonstrations and public pressure from the families of the victims, charges were brought, trials held. On December 27, 1995, Joon Lee, the chairman of Sampoong, was jailed for ten and a half years after being found guilty of criminal negligence. His son, Han-Sang Lee, was jailed for seven years for corruption and accidental homicide. Joon Lee's sentence was reduced to seven years on appeal in April 1996. At the time of writing, Han-Sang Lee's sentence is unchanged.

But, equally crucially, the case led to jail sentences and fines for twelve local officials. Two planning officials from the Socho Ward in Seoul were found guilty of taking bribes. Lee Chung-Woo took the equivalent of nearly seventeen thousand U.S. dollars in 1989 and 1990 to allow design changes and to grant a "provisional use" certificate for the Sampoong department store. Hwang Chol-Min was found guilty of taking a little less for granting final approval for the change of use of the building in 1990.

The Sampoong department store disaster had illustrated once again the potential loopholes in the safety provisions built into structural engineering. Inadequate inspection, change of use permissions, the failure to build to design specifications, had all played their part in the collapse. But there was something new, something more here, too. It was not complacency but corruption, not failure but fraud, that in the end caused the biggest-ever loss of life in a single construction disaster.

III

SAFE AS HOUSES

PROGRESSIVE COLLAPSE, PROGRESSIVE SECURITY

Oklahoma City

IT WOULD BE THE BIGGEST MASS MURDER ON AMERICAN SOIL. A bomb...169 deaths...a building tested to destruction. A test it would fail.

But none of the employees entering the Alfred P. Murrah building in downtown Oklahoma City early on the morning of April 19, 1995 were to know that. The U-shaped, nine-story building, its north façade glazed with full-height windows in a reinforced concrete frame, was a product of its time—designed in 1974 and opened three years later. It was the local working office of a number of federal entities: the Social Security Administration, the General Accounting Office (GAO), Housing and Urban Development (HUD), and a plethora of law enforcement and military agencies among others.

By 9 A.M. many of the employees the building housed were at their desks. Some had left young children at the daycare center on the second floor. One of the early arrivals was Florence Rogers, the head of the Federal Employees' Credit Union, chairing a meeting in her office on the third floor. Another was Luke Franey, a special agent with the Bureau of Alcohol, Tobacco, and Firearms (ATF), who had come in to fix an arrest warrant in his office on the ninth floor.

"I was on the telephone when I heard a loud explosion and I heard the girls in the Drug Enforcement Administration office scream," recalls Franey. An almighty pressure blast then took the ceiling off, toppled a wall and picked up the beefy ATF agent, depositing him more than fifteen feet away from his desk. "I was covered in rubble and I still had the receiver of the phone in my hand with the pigtail hanging off it. The bottom of it was nowhere to be found."

Florence Rogers was even closer to the blast but does not remember any sound, perhaps because of hearing loss. "I felt a tornado-like rush and I was picked up and just thrown against

A computerized simulation of the effect of the blast wave as it shatters the façade of the Alfred P. Murrah building.

the floor." With debris flying everywhere, the eight women perched on the sofas and chairs on the other side of Florence Rogers's desk had just disappeared. "My first thought was they've run off and left me. Then I realized they had not had time to leave." They were among the eighteen employees the Credit Union lost that day.

Within hours of the blast, structural engineers were heading to Oklahoma City on behalf of the Federal Emergency Management Agency (FEMA) office of emergency services. Ironically, the first job of a team in such a situation is to prevent further collapse. The safety of the people trying to get those inside to safety is the priority during any rescue and recovery operation. The consequences of getting it wrong can all too easily compound the original tragedy.

"There have been earthquakes in which more rescuers died than victims," recalls John Osteraas, principal engineer with Failure Analysis Associates (FaAA)—the Californian experts in failure investigation—and chief structural engineer at this new site. "Fortunately there was only one such victim at Oklahoma City." Rebecca Anderson, a thirty-seven-year-old nurse and mother of three followed her instinct to help and rushed into the carcass of the building within minutes of the explosion. She was found wandering around the site dazed and confused shortly afterward and died of head injuries on April 23, four days later.

For hours and days after the explosion, the Alfred P. Murrah building remained a death-trap for rescuers. As they clawed away at what looked like a frozen avalanche of concrete and steel, Mike Shannon, the special operations chief of the Oklahoma City Fire Department, tried to remind his staff of the basics. "You can't argue with that weight of concrete. If you get too emotionally involved you stop working smart....You miss the details, a cracking structure, a creaking column."

Sometimes the dangers meant temporarily abandoning victims. When there was a warning that there was a second bomb on the site Shannon and his team had to walk away from three women entombed on the first floor. "They were screaming, 'For

God's sake don't leave us! You wouldn't leave a dog like this, would you? Why are you leaving?' It was the most gut-wrenching experience of my life, terrible for a third-generation firefighter," says Shannon. "And of course you can't tell them why you're going—that there was another device thirty feet from them—after what they've just experienced."

Having survived the first blast, Luke Franey did find out about the bomb scare. Still on the ninth floor he had managed to get up, go into the hallway outside his office and take three steps before finding he had no way out—the whole of the front of the building had been torn away. Standing on a ledge of floor, he soon saw people running away from the building, getting behind buildings, diving behind brick walls. "I'm telling myself this is obviously not a good sign. Now I don't have anywhere to go, I'm stuck on my little perch."

He had already summoned help from ATF colleagues outside the building by means of a walkie-talkie radio in the office, and could now see the agent he was talking to crouching behind a wall. "I'm getting him on the radio and saying, 'Hey, buddy, what's going on, what's going on?' And this is what he says to me, I'll remember it as long as I live. He said: 'They've found another bomb, they think it's gonna go off. Try and find something to hold onto.'"

Convinced that the building would not stand up to another blast, Franey decided he would rather die trying to get out of it than wait to be buried in another bomb blast. Crossing to the south wall of his floor, he got onto a desk and punched a hole large enough to climb through in a partition wall. He did the same again to get through the next partition wall before managing to cross a fifteen- to twenty-foot outside ledge by hanging onto the broken window frames.

After scrambling over a pile of rubble and getting his first good view down into the interior of the building, Luke Franey found what he was looking for—an intact stairwell. Convinced he had just seconds before another bomb blast, he ran down the stairs only to find himself slipping repeatedly. "Finally after about three floors I looked down and man, I'm just cov-

The precarious state of the debris hanging from inside the remains of the building gives some idea of the danger in which engineers and rescue workers found themselves.

ered in blood, my hands, my feet are soaked in blood from people that had either been carried out or got themselves out after the bombing."

There was no second bomb—it turned out to be a telephone junction box with protruding wires—and with the help of cranes, braces, jackhammers, rescue dogs, ladders, and heavy lifting devices, the rescue efforts were resumed. Dozens of victims were pulled, winched, and lowered from the remains of the building. The trouble was, most of them were dead. What started off as a rescue mission was by the following day a recovery operation. The cause of the collapse—a massive explosion—made it a very different scenario from an earthquake in which buildings tend to topple over, creating voids and air pockets in which people can survive for days. The floors of the Alfred P. Murrah building crushed almost everything as they fell, pancaking to earth. After the walking wounded who left the building by themselves, only eight victims were rescued alive—all within the first eight hours.

The first structural engineers on site quickly established the main problems. There were two main "collapse hazards" as they were known: five damaged internal columns, all still standing but now supporting disproportionate weights, and the east end of the whole building which was now disconnected from the stair/elevator bracing walls and deemed "only marginally stable." There were also six main "falling hazards." They included numerous sections of concrete floor slabs and beams hanging from the edges of the remaining structure; office contents hanging similarly; and badly broken granite veneer panels on the east wall. There were also scores of unreinforced concrete interior wall partitions left leaning dangerously and precast concrete panels which, with their supporting beam connections partly broken on the third floor, were now suspended over the south entry to the building.

Then there was something even more frightening, something everyone could see. It was what became known to rescue workers as the "Slab from Hell" or the "Mother Slab"—a massive section of roof slab hanging perilously from the top of a column on the eighth floor by a single one-inch-diameter rebar or piece of reinforcing steel. "It turned out that rebar had a capacity of about thirty-five thousand pounds so we had something that was right on the verge of failure hanging above the main debris area like the Sword of Damocles," says John Osteraas.

The teams of structural engineers had various options for dealing with the hazards: removal, shoring and bracing, monitoring accompanied by warning systems, and simple avoidance. The trouble was that the whole place was a moving target. The building remained a potentially dynamic force every time something was moved or when something as basic as the elements intervened. Wind and rain were particular hazards. "The debris had to be moved in order, sort of like a game of pick-up sticks where you want to remove each one without disturbing the others to avoid triggering an avalanche or progressive collapse of the debris pile," recalls John Osteraas.

Inevitably it was the "Mother Slab" that attracted the most interest and concern. Dave Hammond of FEMA, the chief structural engineer on site until relieved by John Osteraas, monitored the slab constantly, using a theodolite. The options seemed obvious: cut the single rebar holding the slab or break the slab itself into pieces by normal cutting methods and small explosive charges. The problem was that the consequences of either were like so many of the structural problems at the site—completely unpredictable. One possibility was that if the rebar was cut, the force exerted by the release of such a massive weight could lead to the complete collapse of the east end of the building.

Using explosives to fell the slab was a sensitive issue given that a bomb had caused the disaster and the fact that there were still plenty of bodies to be recovered from the debris. The solution was suggested by a trusted local explosives expert: tie the slab with cables to a stair wall. The cables were secured during the fifth night of the operation and the monitoring was stepped up. "No significant movement was observed," recalls David Hammond, "but the slab did become a constant source of controversy regarding the risk/reward ratio of working under it."

The investigation into the cause and effect of the explosion quickly identified several main sites at the scene. There was the "bite area," as it became known, the huge U-shaped indentation at the front of the building, so-named because it looked as though Godzilla or some other such creature had taken a bite out of the building. There was the pit area, a two-story-tall zone which had been the open atrium but was now filled with debris, all of which had to be removed by hand before dozens of bodies could be extracted. There was "the pile," the main debris area where the nine floors of concrete had pancaked into what looked like a solid mass. Under this was what became known as "the cave"—a void created when a horizontal slab of concrete came to rest on two vertical slabs.

The main problem for the engineers from FEMA and the FaAA was that there were three operations going on at once at the remains of the Alfred P. Murrah building. There was the rescue and recovery operation led by the fire brigade; the criminal investigation led by the FBI; and the structural investigation led by themselves. All three had conflicting interests and priorities yet all three were interdependent. For the structural engineers, making value judgments about forensic engineering in an environment that could not have been further away from the orderly systems of the design office was not always easy.

From an investigative point of view, the first and overriding question for the engineers was why one bomb had caused so much obvious damage, taking nearly half the building down. The building was, after all, solid, being built of cast-in-place, reinforced concrete, the basic material of military bunkers. It had conventionally reinforced columns, girders, beams, and slab-bands. Although it had not been built to resist earthquakes it was constructed to withstand tornadoes, a common hazard in Oklahoma. The required wind-load resistance was met by reinforced concrete walls, with the required resistance rising progressively from twenty-five pounds per square foot up

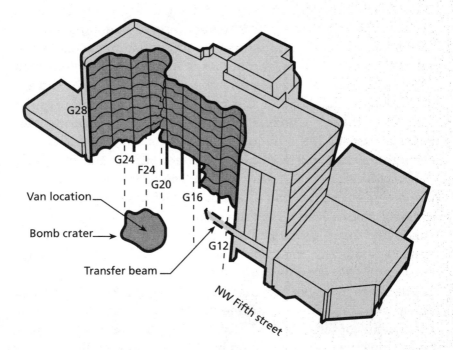

FIGURE 6
The Alfred P. Murrah building after the blast.

to heights of thirty feet to forty-five pounds per square foot for heights over one hundred feet.

The scale of the destruction and the apparent mystery made it an unprecedented learning opportunity for U.S. structural engineers. The building had in effect been "tested to destruction"—a rare phenomenon in the construction trade. Cars, aircraft, high-value consumer goods: all of these go through a rigorous design and testing process when prototypes are rolled, crashed, and crushed. Weaknesses and flaws are understood, counteracted, eliminated. All buildings are, however, one of a kind, a prototype in themselves. "It's never tested until we have an extreme event like an earthquake or a tornado or an explosion," explains John Osteraas. "It is from those rare catastrophic failures that most progress has been made in engineering...that's the core reason for studying our failures—to be able to change practice...to eliminate the mistakes and misconceptions of the past."

The cause of all the damage was soon clear: it was a vehicle bomb. The crater, centered just ten feet away from the center of the north side of the building in NW Fifth Street, was some twenty-eight feet across and nearly seven feet deep, with more than six thousand cubic feet of soil and asphalt. Part of the axle deposited nearby soon gave the FBI a VIN, a vehicle identification number, enabling agents to identify a Ryder truck rented in Junction City, Kansas.

The proximity of the detonation to the front of the building was the first obvious reason for the extent of the damage. Curbside parking was not only allowed here, there was an indented passenger-loading zone just seven and a half feet from the exterior. But even more crucial, engineers discovered, was the proximity of the vehicle to one of the columns that supported the entrance to an open pedestrian plaza. Above this, the north façade was glazed with a full-height window/wall cladding, designed to optimize natural light.

The key column, designated G-20 on the plans, was one of four spaced forty feet apart, two stories high (see Figure 6). The remains of the G-20 column showed that it was the only one of the four with direct air-blast effects from the explosion, yet all three of the other columns—G-24, G-16, and G-12—had also been felled in the disaster. The engineers in Oklahoma were looking at the same thing that had faced investigators at Ronan Point in London years earlier (see Chapter 7): disproportionate damage. In fact, in Oklahoma it was even more disproportionate. At Ronan Point there had been what engineers call a vertical propagation of the collapse—one floor had gone and twenty-one others had followed. In Oklahoma it was both vertical and lateral. All the floors above the G-20 column had collapsed but so had all the columns to the side.

Why had there been a progressive collapse? Engineers talk of structural "redundancy"—had there been a loss of redundancy here? What would redundancy have meant in the Alfred P. Murrah building in Oklahoma?

"The best way to describe redundancy is the old belts and suspenders analogy. You put on a belt and you put on suspenders just in case," explains John Osteraas. "When we design, for example, aircraft systems, there are four parallel systems so that three of them can fail and you still have a fourth system that will allow you to control the airplane. It's similar with a structure. You can provide multiple paths where the loads in the building can find their way to the foundation. It's those multiple paths, those alternative paths if you will, that give you what we call your redundancy."

A report on the Oklahoma City incident jointly published by FEMA and the American Society of Civil Engineers in August 1996 put it very succinctly. "Redundancy is a key design feature for the prevention of progressive collapse. There should be no single critical element whose failure would start a chain reaction of successive failure that would take down a building."

Studying the design, it soon became clear to Osteraas that the shattered building was "the antithesis of redundancy." With outline designs of the Alfred P. Murrah building on his computer screen in front of him in his office at Menlo Park, he demonstrates how there was a loss of redundancy as loads were transferred to the ground. "In the upper stories we had ten columns across the north face of the building," he says, pointing to the top part of the building. "At the third floor there was a big transfer girder which transferred these loads onto only four columns."

The consequences were obvious and are clearly traceable on Osteraas's computer model. Below the third floor, one column—say G-20—was taking the load of three columns above. "On the upper stories you can see the area in red here is all of the area of the building supported by a single column," he says, pointing to a swathe of the building highlighted on the screen. "So if we lose G-20, everything shown in red halfway up the building is going to collapse."

As it did. Osteraas's computer sequence shows what happens in the first few milliseconds—less time than it takes to blink—that followed the blast as the wave moves out in a hemisphere from the Ryder truck. The initial damage was a complete shattering of column G-20. "This process is known as brisance," explains Osteraas. "The shock waves move through and reduce solid concrete literally to sand and gravel. It's pulverized."

THE BLAST WAVE MOVED UP AND THROUGH THE BUILDING AT SUPERSONIC speed. Within fifty milliseconds, the concrete floor slabs on the third, fourth, and fifth floors—never designed to carry upward loads—bowed upwards. The pressure of the blast also caused cracking of the connections between the floor beams and the interior columns. "It's this cracking that leads to the ultimate failure of the floor slabs, because once the blast wave passes and gravity takes over this connection is no longer able to support the dead weight of the slabs," notes John Osteraas.

Within seconds the whole north façade of the building was pancaking. Again the force was startling. "When you looked at the underside of the floors you could see that the reinforcing bars had been ripped out," says Eve Hinman, a senior engineer at FaAA who joined Osteraas at the Oklahoma site as a blast consultant and specialist in defensive design. "It was clear that the building came down as the floors came down, it was ripping out these reinforcing bars."

An aerial view of the scale of the devastation of the Alfred P. Murrah building in Oklahoma City—a bite-shaped chunk was torn out of the north side of the building by the four-thousand-pound bomb.

But if that explained the vertical progressive collapse, what about the lateral collapse, the toppling of columns G-12, G-16, and G-24? The first clue was the transfer beam that had straddled them all. It was found lying intact upside-down, having toppled inward and taken the three columns with it. Although there were some shear lines on the transfer beam, there was no break or failure. "This was not a direct air-blast effect, this was progressive collapse," concludes Eve Hinman.

Again the computer model makes clear what would have happened once column G-20 had been pulverized. With the columns forty feet apart, the transfer beam was now expected to straddle an eighty-foot gap without support. That was simply too much. Without adequate support, with the air-blast applying pressure from the outside and later the tumbling floor slabs applying weight on the top, the five-foot-thick transfer beam, the key to the support of the whole building, tumbled backwards into the remains of the structure.

But examining the transfer beam investigators discovered something else—another reason it could not have stayed in place. It was not connected to the four key columns by steel reinforcing bars inside the reinforced concrete. "It wasn't monolithic. You want it to be a solid, heavy building that's not going to go anywhere, where all the elements are tied together," notes Eve Hinman. "They just weren't in the Murrah building." As in Britain at the time of the Ronan Point disaster, such "continuity" was not obligatory under the building regulations.

Progressive collapse can by the very nature of being progressive leave a tell-tale trail of clues—a sequence—through a collapsed building. As the extent of the failure and damage diminished further away from the epicenter of the blast, the failure detectives were able to pick up important indications of how joints and materials had reacted in areas which were inaccessible or totally destroyed. These damaged features—particularly connections between beams and columns—were essentially frozen in failure at the moment of the blast.

SEQUENCING IS THE KEY TALE TO BE READ IN THE RUNES OF THE CONcrete. One good example was F-24, an interior column whose collapse created the bite-shaped area. The column was found intact, on top of the second- and third-story floor slabs. This proved conclusively that the column had not failed as a result of the direct effects of the blast but had fallen after the floor slabs above it had collapsed. "It's very much like an archaeological dig," says John Osteraas. "You excavate down through the layers and you can go back through time, back through the sequence of failure."

The effects on the building and the detective work done elsewhere soon enabled the engineers to calculate the size of the bomb at about four thousand pounds of ammonium nitrate—

agricultural fertilizer—and fuel oil. After shattering the metal sides of the van—the largest piece of metal found was eighteen inches long—the blast wave that was generated would have slammed into the federal building at a force of nearly six thousand pounds per square inch. But the force did not dissipate as it hit the next obstacle, the façade of the building. Initially it was reflected backward, then amplified. "Just as a sea wall causes an on-rushing ocean swell to reverse direction and then jump higher up the wall, the building momentarily exerts this effect on the shock wave," notes Eve Hinman. "That wave would have caused the building to sway as if in an earthquake measuring 5.0 on the Richter scale."

But something else also magnified the effect of the blast wave: the glass windows on the northern façade of the building. As they shattered, thousands of pounds of glass shrapnel were carried into the building by the air blast as it rushed upwards. Back at ground level, air rushed in to fill the partial vacuum created by the blast. "This rushing air mass generated a force a thousand times stronger than the strongest hurricane," notes Eve Hinman. "It was this wind that carried debris everywhere, exerting massive pressures on all the surfaces it encountered." Windows up to one thousand feet away from the origin of the blast were blown out.

The blast wave's entrance into the building was further facilitated by the incorporation of the pedestrian plaza into the design. The two-story open area adjacent to the G-20 column where the van was parked had acted like a funnel—or even the kind of wind tunnel in which some of the experiments to learn more about the collapse of the building were conducted. "There was almost a natural cave effect that allowed the blast wave to enter the building easily and apply pressure to the underside of the floor slabs on the lower stories," observes John Osteraas. "It couldn't have been better... or worse, depending on your perspective."

The whole collapse sequence took just five or six seconds as the domino effect of progressive collapse took hold. Incredibly, the whole sequence was recorded on an audiotape being used to document a meeting which began at 9 A.M. in the water resources building, just one block away from the scene. The noise attributed to the explosion is less than half a second, followed by about a second and a half of blast wave, then about three and a quarter seconds of deafening crash—a noise that exceeds the range of the recorder—as progressive structural collapse follows. "In effect, the building was irretrievably damaged fifty milliseconds after the blast," notes John Osteraas. "After that the domino effect and gravity just took over."

Despite the glass, the pedestrian plaza, the proximity of car parking, the poorly supported columns and transfer beams, and the absence of reinforcing rods tying the two together, some design features worked well in the Alfred P. Murrah building. Indeed, these features probably saved lives. The four air shafts at the corners of the building and the distance of the elevator shafts from the loading bay where the bomb exploded all helped. The poured-on-site reinforced concrete used in construction is heavy and tends to deform rather than break in response to excessive loading. Had the building been precast concrete the damage would have been infinitely worse. Laboratory analysis of samples of both the concrete and steel used in the Oklahoma City federal building showed it was well above the specifications required.

The key lesson from Oklahoma was the importance of structural redundancy, building in fail-safe mechanisms that make disasters like this more survivable for those caught inside. "The key design concept is one of life safety. We're willing to sacrifice the building in terms of extensive damage if we can keep it from collapsing," says John Osteraas. "An automobile might be a total loss but as long as you keep the passenger compartment intact you have protected the occupants and given them the chance of escape without serious injury. We can apply the same principles to a building to prevent collapse. We need a predictable, desirable mode of failure when the building is overloaded rather than a random, chaotic failure."

But there was another lesson in the Oklahoma City bombing: follow-up after such a major structural disaster is crucial. The research, design development, and best practice that follow such an investigation has to be written into building codes if it is to prevent future collapses. "After the Ronan Point [disaster in 1968], we saw a lot of funding being made available to research progressive collapse and to start coming up with measures that would stop that kind of collapse from happening," recalls Eve Hinman. "Unfortunately, after a few years that funding dried up. What I found in the research I did on progressive collapse is that a lot never went beyond the research stage. Most of the measures are not explicitly in the codes as we see them now."

Many of the basic design and construction measures compulsory in U.S. earthquake zones would have saved lives in Oklahoma City. The issue, particularly in the construction of government buildings where budgets are always tight, is cost. "Seismic design in building increases the structural design cost by 20 percent and the entire construction cost by 2 to 4 percent," says Eve Hinman. "The contract can specify that the building should be designed for progressive collapse. That obliges the contractor to design the building to remain standing even if one column is removed."

In August 1996, FEMA and the American Society of Civil Engineers (ASCE) published a report on the Oklahoma City bombing entitled "Improving Building Performance through Multi-Hazard Mitigation." The report looked in detail at compartmentalized construction, a system known as special moment frames, and dual systems with special moment frames—all systems used in earthquake zones but whose emphasis on reinforcement, connection, and ductility were critical in the face of a blast. None of these systems was available until 1985, ten years after the Alfred P. Murrah building was completed.

It is, of course, hypothetical yet fascinating to estimate what the damage to the building would have been if special moment frames—the most recent and most effective earthquake-resistant design—had been used. The FEMA/ASCE report did just that. In all probability, the G-20 column would still have been destroyed by brisance, but if it had survived the loss would have been limited to those floor slabs destroyed by the actual air-blast. That would have reduced the loss of floor space by as much as 85 percent. If G-20 had not survived the blast, losses would have been reduced by only 80 percent. "Even though individual columns and slabs would not have had enough strength to avoid being cracked, the reinforcing steel would have held many of the building elements in place, keeping large portions of the building erect (at least sufficiently erect to allow the occupants to escape after the blast)," the report concluded.

Eve Hinman and David Hammond wrote their own report, "Lessons from the Oklahoma City Bombing, Defensive Design Techniques," which was published by ASCE. Eve Hinman in particular drew on her specialist experience—designing nuclear missile silos, NATO military facilities, industrial buildings exposed to accidental explosion, and civilian buildings vulnerable to terrorist attack. The report highlighted three main areas of concern: one, to build better-protected new buildings; two, to take remedial action to better protect those built before better protection was available or considered necessary; and three, to see any structural measures as part of a whole security system—including a building's access, design, operational procedures, and evacuation plan.

Eve Hinman says it boils down to three things. Do everything you can to reduce the car- or truck-bomb threat by keeping vehicles away from buildings, as U.S. embassies were doing some time before the Oklahoma incident; make the exterior of the building as blast-resistant as possible, paying particular attention to its shape and the glass used; and make the structure of the building as robust as possible with precast reinforced concrete and steel the key materials. There is also much to be done on interior design and training those who use the building. "Maybe you shouldn't put your computer against the window, because when the blast comes through the window, the computer will be blown against you," she says. "Things like that can mean the difference between being severely injured and walking away from the incident."

Eve Hinman advised the Port Authority of New York and New Jersey after the attack on the World Trade Center in 1993. Two other major incidents—the bombing of the Khobar Towers in Saudi Arabia and the deadly attacks on two U.S. embassies in Africa—as well as the Oklahoma bombing have kept federal attention focused on building security and survivability. Federal building codes have been changed as a result of the recommendations of a committee she sat on. Redundancy of structure, continuity of joints and reinforcing steel, and ductility of all materials are now basic to what is seen as a more "belt and suspenders" approach to structural integrity. Beams must now go in two directions, and interior walls must effectively act as thick beams so that they can support a downward load without collapsing if the floor slab is lost.

But there is a long way to go outside the federal building market. In 1998, a survey showed that 68 percent of building design professionals in the United States had not changed their design processes or considerations in response to the recent high-profile disasters. The problems are complacency, cost-consciousness, and "building-code mindset" according to Rudy Matalucci, project leader of Architectural Surety, a scheme dedicated to applying the "surety" principles required in nuclear-weapons work to the problems of making structures safer. On existing buildings the situation may be even worse. "I read in *The New York Times* that 88 percent of the U.S. embassies abroad do not meet the one-hundred-foot setback requirement [for vehicles]," notes Eve Hinman. "That should give you an idea of how few buildings are up to spec."

Indeed, for those who worked on the Oklahoma City site, the horror of what they experienced is enough incentive to continue propagating the gospel of better buildings. Eve Hinman and John Osteraas are just two of the structural engineers who need no reminding that the actual bomb killed very few of the victims of the Alfred P. Murrah building. The collapse itself accounted for the vast majority. "The real story here is the human side of things. It's not the structural," says John Osteraas. Eve Hinman concludes, "Although I'd worked on explosive effects for years I had never really been exposed to the direct after-math of one of these events. What I carried away from that incident is a better understanding that our primary objective is to protect the occupants—that is my focus now."

FIVE WEEKS AFTER THE BOMBING THERE WAS ANOTHER EXPLOSION IN downtown Oklahoma City. On May 23, 1995, after intense debate about whether the ruins of the Alfred P. Murrah building were a suitable monument or a brutal reminder, 150 pounds of dynamite strategically packed in 300 locations around the ruins were detonated. Two thousand people turned up to watch, including the families of Christy Rosas and Virginia Thompson, whose remains were still entombed inside.

Within eight seconds, the Alfred P. Murrah building was no more. A few minutes later the clearance teams were at work on another pile of rubble. The investigation was over, the lessons evaluated. The only question for the years ahead was: would that knowledge be applied?

7

FAULTY TOWERS

Ronan Point

CAROLE EUSTACE HAD NOT WANTED TO MOVE BUT ON BEING SHOWN around the new twelfth-floor, one-bedroom apartment, had quickly changed her mind. The living room was large and one wall was completely glazed, giving magnificent views over the River Thames. There was underfloor heating and with gas mains rather than electricity the apartment would be cheaper to run. "We just fell in love with it...you just couldn't say no," she enthuses.

Indeed, no one seemed to say no. By May 16, 1968, just a few weeks after Carole Eustace was shown her apartment, all but eight in the block were occupied. But just before six o'clock that morning, still in bed, her dreams were shattered. "I heard this massive bang, then all of a sudden my husband threw himself on top of me and said the apartments are falling down...we're going for a ride," she recalls. "The noise continued and seemed to get louder...and as I got out of bed the wall was missing behind my bed." In the hall Carole and her husband discovered something else that had gone. "As my husband opened the living room door it just fell off. There was nothing there whatsoever...from the end of my hall there was no living room."

Others had similar close shaves. "The whole place shook. Suddenly our bedroom wall fell away with a terrible ripping sound. We found ourselves staring out over London, our heads just two feet away from the eighty-foot drop," recalls James Chambers, who was in bed with his wife Beatrice in their seventh-floor apartment. "Showers of debris and furniture were plunging past us. We heard screams. I think it must have been someone falling with the debris."

Brenda Maughan had been in her living room, one floor above the Eustaces' apartment. Unable to sleep and not wanting to disturb her husband, she had left their bedroom at about 5 A.M., eventually dozing off on the sofa. Suddenly virtually the whole room slid into the early morning light and Brenda found herself hanging on to a narrow ledge, her feet and legs covered with rubble. Eventually her husband was able to get an arm through the blocked living room/hall door and grab hold of his wife. With the other hand he gradually cleared away enough of the rubble blocking the door to pry it open and pull Brenda to safety. Chipped teeth, a dislocated shoulder, and a broken leg were diagnosed in the hospital, but she was alive.

A full view of what happened was only possible from outside. One witness in a nearby factory heard an explosion and looked up at the twenty-two-story tower block, only to see first one entire wall then another, then another popping out of the building, like slices of bread being forced out of a toaster. Sideways then downward they fell, followed by the now unsupported walls, floors, and ceilings they left behind. Gradually the whole south-east corner of the block collapsed, floor by floor. Ronan Point, the towering pride of Newham Borough Council's commitment to a massive housing program, a system-built block erected in just eighteen months and occupied for less than ten weeks, had collapsed like a pack of cards.

THE COLLAPSE OF RONAN POINT WAS TO BE PIVOTAL IN THE DESIGN AND construction of social housing in Britain. The progressive collapse of the twenty-two floors was followed by a rolling campaign that over the next seventeen years was to expose the inherent flaws in system-built high-rises. It was a tale that played a key role in changing the public perception of high-rise apartments from irresistible "homes in the sky" to decaying death-traps.

It did not take long to establish that there had been an explosion at Ronan Point. Just after 5:45 A.M., Miss Ivy Hodge, a fifty-seven-year-old cookie and cake decorator who had moved into her eighteenth-floor one-bedroom apartment only four weeks earlier, had gone into her kitchen to make a cup of tea. She filled a kettle, lit a match, and remembers being "thrown to the ground." Coming to in a puddle of water some minutes later, she managed to make her way out of the apartment and with the help of neighbors staggered down seventeen flights of stairs.

As Ivy Hodge left for the hospital for treatment for burns and shock, firemen soon discovered that some of her neighbors had not been so lucky. With lookouts warning of any movement in the partially demolished floors, they used cranes and lifting gear to uncover four bodies. Thomas and Pauline Murrell had lived in Apartment 110 on the top floor, the twenty-second, and Thomas McCluskey and Edith Bridgstock had lived in Apartment 85 on the seventeenth floor, directly below the site of the explosion. They had all died of multiple crushing injuries. A fifth occupant, Anne Carter, eighty-two, died in the hospital two weeks later.

Yet one look at the floor slabs and the debris flapping in the wind was enough to demonstrate how much worse it could have been. This was a progressive collapse through all twenty-two floors, a collapse that had sliced off the living rooms of all the one-bedroom apartments in the southeast corner of the block. Because it all happened so early in the morning, the dozing Brenda Maughan was the only resident in a living room. Indeed, the block was so new that tenants were still being moved in. Of the eight empty apartments, four were in the southeast corner. Of the four apartments above that of Ivy Hodge, only one was occupied, the one housing Thomas and Pauline Murrell, two of the four victims.

Within hours the newspapers were reflecting public outrage. "WHY? WHY? WHY?" ran one headline. "TOWER OF TERROR" proclaimed another. All sported aerial photographs of the block, whole floors and ceilings hanging down from the top four floors like limp pieces of soggy white bread, the remainder a pile of assorted debris at the foot of the block. Sam Webb, a young junior architect in the London Borough of Camden's housing department, recalled the sense of shock in his office. "Someone held up the midday paper with the pictures and you could have heard a pin drop. Everyone stopped. Somebody said: 'A friend of mine is designing one of those.'"

Nowhere were there more of such tower blocks than in Newham. Ronan Point was just the second of six, four of which were already completed. They were built by means of the Larsen–Nielsen system, developed in Denmark in 1948 and since then licensed across the world. The system was never intended for more than six floors when originally developed, but engineers and architects had resorted to the old temptation of making a good idea better by tampering with it.

Essentially, the walls and floors were prefabricated using precast concrete, then lowered into place on site. With such panel buildings there is no steel frame; no lateral girders. The principle is simple: the floors and walls *are* the frame. The apartments were, as Sydney Lenssen of *Construction News* put it at the time, a series of boxes, one on top of the other. The walls support the floor above, that floor supports the walls above, the same principle as a house of cards. The building's stability comes from the dead weight carried down through the crosswalls. Gravity essentially keeps them standing.

Inevitably, the joints and the accuracy of fitting assume a very special importance with such off-the-shelf building systems. Swedish and Danish constructors worked to a tolerance of no more than one-twenty-fifth of an inch in measurements of several yards. Sam Webb describes fitting on site as needing the precision of watchmaking. "Whoever had designed them hadn't really considered what it was like working 180 feet above the ground, on a wet Friday afternoon, when there's no scaffolding or safety net and you've got to kneel on your hands and knees and make a very narrow concrete joint."

But any doubts were banished by the optimism and hope engendered by such an apparently magic solution. In the mid-1960s, high-density high-rises seemed to be the answer to all east London's housing problems—slums, wartime bombing, and a growing population. Industrial housing units were cheaper to buy, quicker to erect and, above all, avoided the need for the skilled building labor that was in such short supply at the time. In the end, the key advantage was the key flaw.

The public inquiry, announced the same day as the disaster by a startled government, ran until early August. Headed by a past president of the Institute of Structural Engineers, Sir Alfred Pugsley, the inquiry took evidence from more than two hundred witnesses and rescue workers, with assistance from a phalanx of experts and research teams. The inquiry quickly established the cause of the explosion—domestic gas. The cause of the leak was also quickly established: a substandard fractured brass nut that connected a flexible hose at the back of the stove to the standpipe that supplied gas to Ivy Hodge's apartment.

Quite quickly it became clear that the initial explosion had not amounted to much. It had not damaged Ivy Hodge's eardrums, suggesting that the pressure was less than ten pounds per square inch. Three biscuit tins recovered from the kitchen were charred and buckled—tests indicated that they had been subject to pressures of three to nine pounds per square inch. Tests also showed that what had taken place at Ronan Point was a multiple explosion which had been magnified as it spread through the open spaces of Miss Hodge's apartment, igniting gas that had collected in rooms and corridors. In essence, the force of the explosion had been weakest at the point of the ignition, strongest on the external walls.

All this had a devastating impact on the H-2 flank wall joint where the floor and wall were simply bolted together. The inquiry commissioned tests from the Building Research Station and Imperial College London and found there was almost no resistance to an internal explosion. Indeed, the tests showed that the wall panel would be blown out whole at an average pressure of just three pounds per square inch, perhaps less than a third of that of the actual explosion. Ironically, had the explosion occurred on a lower floor, the extra friction resulting from the additional loads compressing the wall panels might have actually prevented the disaster.

Once one panel wall was blown out, those above it were unsupported and collapsed. They in turn crashed onto the floors below, causing the "progressive collapse." That brought the whole southeast corner of the tower block down. At this point a building's structural "redundancy"—its ability to carry loads by more than one mechanism if something critical fails (in other words, its fail-safe margin)—comes into play. The trouble was, Ronan Point had none.

It was not as if the concept of structural redundancy was new in engineering. Designing to ensure that a local crack could not lead to a general collapse was standard—but not in system-building, where arguably it was more essential than ever. The inquiry report was measured but adamant on the matter of redundancy. "It is unfortunate that among the few structural engineers who have been concerned with system-building in this country, few indeed seem to have given thought to this aspect of structural design. In the case of Ronan Point the specification certainly did not touch on the matter."

Worse still, in making no allowance for structural redundancy, Ronan Point's design and construction appeared on the face of it to break no rules. Firstly, there was no code of practice relating specifically to large concrete panel construction. Secondly, although the building regulations of the time contained a "catch-all" clause called a functional requirement on structure, they took no specific account of structural redundancy and progressive collapse. Ronan Point fell literally and figuratively through the gaps in practice and regulation. System-building was not only too new for the architects and engineers of Britain, it was too new for the regulators and politicians.

Over the years, many have accused the inquiry of focusing far too closely on the cause of the accident and too little on the effect. Critics complained that the originator of the Larsen-Nielsen system had not been called to give evidence or to say why the system was not used above six stories in Denmark without major redesign, and that the inquiry had omitted to cross-examine the engineer who designed Ronan Point. Yet the seventy-page report, published six months after the disaster, did raise critical concerns, often boldly straying from its remit to ring warning bells on key design flaws. How could anyone ignore the following?

1 Wind

Ronan Point had been designed to withstand only the pressure exerted by a wind speed of 63 miles per hour. The inquiry determined that there was good evidence that at two hundred feet above ground level a wind speed of 105 miles per hour could be expected to occur every sixty years—the intended lifespan of the block. The suction effect of the pressures applied by such winds, in particular the opening of the joints as the tower block bent in the wind, could have a similar effect to the explosion in Ivy Hodge's apartment. Whole wall panels could be torn out at the top corners of the building, leading to progressive collapse. The inquiry urgently recommended strengthening the face panels, flank wall panels, and joints in the load-bearing walls. By implication it also recommended updating the building codes on wind loading. The codes were fifteen years out of date as to the wind forces such a building might have to withstand.

2 Fire

The impact of the heat of a fire could have a similar effect to wind on the joints—it might move them, the inquiry decided. Although the individual components of the building provided the specific fire resistance necessary, the overall design might not. "It is estimated that a fire could so expand and 'arch' the floor slab and bend the wall panel, as to displace or rotate an H-2 joint to a dangerous degree. It seems essential that this possibility should be studied in any modification of the H-2

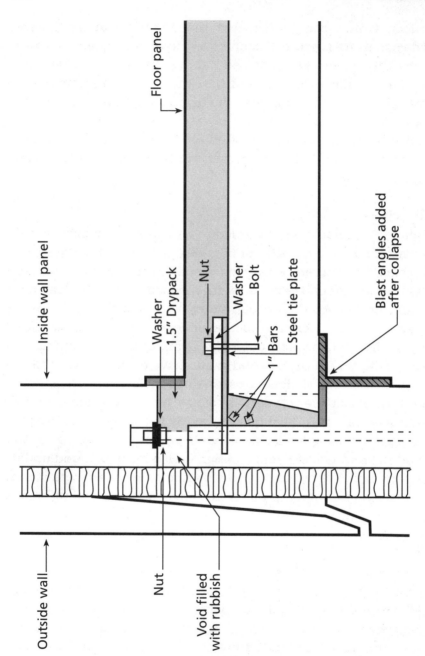

FIGURE 7
Detail of an H-2 flank wall joint.

joints," the inquiry concluded. (See Figure 7 for detail of an H-2 flank wall joint.)

3   Joints and construction

It was not just the design of the building that worried the inquiry. The report raised several questions about the quality of the construction, in particular that of the joints on which the building's overall safety depended. The inquiry had one H-2 flank wall joint opened up and found that the drypack mortar that secured the wall and floor panels could only be rammed into the one-and-a-half-inch gap from inside the building (see Figure 7). Because it was impossible to ensure that the packing was as good at the outer face of the joint as at the inner, some of the mortar on the outer edge of the inspected joint was loose. Moreover, the drypack mortar itself was mixed to the wrong specification—one part cement to one part sand instead of one part cement to two of sand.

4   Tie-plates

There was even more evidence of negligence with the steel tie-plates that connected the walls to the floor slabs. The stud used was supposed to pass through an oval slot in the tie-plate and screw into an insert put into the floor slab when it was cast (see Figure 7). "We had some fifteen of these tie-plates inspected at Ronan Point and in all cases there was evidence of poor workmanship, mainly by way of failure to tighten the nut of the stud down so as to press the plate against the surface of the floor slab," the inquiry reported.

THE INQUIRY CONCLUDED THAT NONE OF THESE DEFICIENCIES IN EITHER workmanship or supervision contributed to the disaster—the subject of the report. But the discoveries raised as crucial a range of questions as the original disaster had. Why might these deficiencies, collectively or singly, not contribute to another disaster in the future? Were Ronan Point and its sister blocks quite literally houses of cards waiting to be felled by the right wind, the wrong fire or the failure of their own creaking joints?

One man who was convinced the place was a potential death-trap was Sam Webb. He had attended the inquiry and took a detailed interest in its findings. He became even more convinced when, despite eighteen specific recommendations from the inquiry panel, remedial action was minimal. Within a year of the disaster the debris had been cleared away and the southeast corner of Ronan Point rebuilt as a separate one-apartment-wide reinforced concrete tower that on completion was tied in to the main building. The rest of the structure was reinforced with blast angles, broad metal plates in the shape of right-angles which, when attached to the joints, tied the walls and the floors together. The government agreed to a dual standard for system-built apartments, with all new structures being built to resist explosions exerting pressures of up to five pounds per square inch but existing ones having to be strengthened to withstand only two and a half pounds per square inch. With the gas in system-built apartments in Newham now disconnected, this level of resistance to explosive force was considered sufficient.

The prevailing attitude seemed to have been reflected by Newham's barrister, Desmond Wright QC, during the inquiry. "The architect himself feels that this block is perfectly safe; he would probably say 'as safe as houses'—but you will understand what I mean."

Declaring the blocks unsafe or unfit for occupation was just not an option. Public alarm was already high enough—it had to be quelled, not fuelled. The nearly five million pounds borrowed to build the series of Larsen–Nielsen blocks that included Ronan Point had assumed a lifespan for the blocks of sixty years. Although Ronan Point was only the second in the series, four more had now been finished in Newham alone, two of them while the inquiry was sitting. Ironically, because work on system-built tower blocks had not stopped during the inquiry, the political and economic stakes to declare them safe were much higher once the report, with its damning findings, had been published. By the time the inquiry's findings were published in November 1968, the number of Britons housed in system-built tower blocks had doubled.

Having to rehouse these tenants—let alone strengthen or even destroy new homes at a time when slum clearance was a national priority and the government had made ambitious manifesto commitments—was a frightening prospect. "The politics were just insurmountable, the pressure on expert witnesses extraordinary," recalls Sam Webb. In no single person were the political pressures better exemplified than in Elwyn Jones. He was the Labor government's attorney-general and as such the man responsible for setting up the inquiry. But he was also the Member of Parliament for the part of Newham where Ronan Point had collapsed.

In February 1969, many of those critical of system-building and the inquiry into Ronan Point attended a public meeting at City University in Central London. Sam Webb submitted a paper, later published in the *Journal of the Institute of Structural Engineers*, predicting serious problems with the joints at Ronan Point. Sir Alfred Pugsley attended to address the audience in what seems to have been the only time he spoke publicly about Ronan Point. "Pugsley mentioned all the technical submissions to the inquiry, said these had not been printed but were available to those who wished to see them," recalls Sam Webb. "He seemed to go out of his way to encourage people to look at them."

As he had said quite unequivocally in his report, Sir Alfred seemed to be saying there was a lot more here that merited further investigation. Amid growing reports of problems with system-building elsewhere, Sam Webb decided to take up the offer and put the system to the test. Would all the raw documentation submitted to the inquiry be accessible to the public? In March 1970 he knocked on the door of the Ministry of Housing and Local Government in Marsham Street, London, and asked to see all the files on Ronan Point. Somewhat to his surprise no one told him to get lost. For more than three weeks he sat at a large desk and studied the transcripts, submissions, and notes. He could not make photocopies, but he could take notes. He did, copiously, as he uncovered a raft of interesting documents.

Item one was a couple of deletions from the section of the report on the structure of Ronan Point written by Sir Alfred Pugsley. In his draft Sir Alfred had written: "However, in popular terms, to make walls strong enough to resist sixty pounds per square foot is only to make them about as strong as the glass in a good window." The sentence was omitted from the final report. It was by no means the only such deletion. A paragraph describing the strength of Ronan Point's walls and joints (in being capable of withstanding pressures of forty-five pounds per square foot) as being "about as strong as the glass in ordinary windows" had also being omitted. The warnings about the effect of fire on the joints, quoted on page 143, seemed to have been diluted.

As regards the effect of internal pressure on walls and joints, Sir Alfred had apparently written an alternative in the margin: "However, it would seem to us very unfortunate if in a wind liable to break many glass windows (those at Ronan Point have been found by test to break at sixteen pounds per square foot) the inhabitants at Ronan Point should have to worry also about its structural stability." Even that sentence did not appear in the final report. It was a particularly apposite metaphor, according to Sam Webb. "What he meant by that was that if someone inadvertently left windows open at the top in a corner, or if a wind of say eighty to one hundred miles per hour was strong enough to break the glass, then the effect of the pressure on the load-bearing flank walls would be sufficient to cause them to move."

Item two was a letter dated August 1968 from A. D. F. Gilbert, the solicitor for Taylor Woodrow-Anglian Ltd. who held the British license for the Larsen–Nielsen system. It referred to a group of critical architects and engineers, including Sam Webb, Bill Frischmann, George Fairweather, and Bernard Clark, who had raised doubts, concerns, and broader issues about the whole manner of system-building and the way it was being applied in Britain. "The letter was pretty clear in my opinion—the gist of it was how do we shut these people up," recalls Sam Webb.

These revelations seem to indicate a possible struggle that had gone on behind the scenes at the inquiry. In 1971, armed with such evidence—more than fifty pages of notes and quotes—Sam Webb enlisted the help of the Labor MP Tom Driberg. Driberg went and saw Elwyn Jones, by now in opposition under a Conservative government. "He was supposed to come down and talk to us but he refused. He just sent Tom down with a message that if I repeated any of this outside the Houses of Parliament I'd be, in his words, done for criminal libel," recalls Webb. "It was an impossible position. At that time I had no access to the documents other than the notes I had made."

Having decided he could do more out of prison than in, Sam Webb bided his time. Fourteen years later he felt he was vindicated. In October 1984, amid intense concern about system-built tower blocks and under pressure from the media and Parliament, in particular from local MP Nigel Spearing, Ian Gow, the Minister for the Environment responsible for housing, ordered the release of all the documentation from the inquiry. In February 1985 Sam Webb, accompanied by Nigel Spearing, found himself back at Marsham Street, facing a long table covered in files and an assortment of civil servants. "They were all lined up, apparently by rank," reminisces Webb. "It was like something out of the television sitcom *Yes, Minister.*"

Webb immediately asked by reference numbers for some of the documentation he wanted to show Spearing. "The most senior official turned to the next and so on right down the line," says Webb. "Finally the message came back: it wasn't filed like that any more. They had changed their filing system." And lost some paperwork in the process, it seemed. Further searching would reveal that although the differences between the draft and the final report, and Sir Alfred's handwritten alternatives in the margins on the subjects of strength and fire, were there for all to see, much else was not.

The department finally admitted that it had re-ordered and refiled the documentation in 1984 and had lost what it referred to as H4/758/3 Plan Covers 14–27, some fourteen major files. It was an awful lot of paper—half the documents submitted to the inquiry—all lost or destroyed some time between March 1970 and February 1985. Today, the index of the paperwork on the Ronan Point Inquiry notes that "certain of the papers in those missing plan covers have been located." In being "properly ordered, secured, and catalogued," as Ian Gow announced, the evidence had, Webb believes, been censored.

In the fourteen years it took to get what was left of the Ronan Point Inquiry documents released, concerns about the social impact of tower blocks heightened the safety concerns. In 1983, Sam Webb was asked to talk to the First National Tower Blocks Conference about Ronan Point. Sue McDowell, the leader of Newham Tower Block Tenants' Campaign, heard him speak. She had formed the group two years earlier after a friend of hers had committed suicide by throwing herself from the twenty-second floor of their apartment block in Stratford, also in Newham, East London.

Sam Webb learned from tenants that Newham was refurbishing the whole estate on which Ronan Point was located. In a brief talk he told them what he thought would be happening to the building fifteen years after construction, given what he knew. "I told them there would be gaps between walls and floors through which smoke would pass; that you'd be able to hear people and their television on different floors," Webb recalls.

The talk struck a chord with several tenants: Webb had described the reality in which they were living. They invited him to come over. "One of the simplest tests was to get a sheet of paper, tear a strip off, put it against the skirting board, and let it go at one end. The loose end was coming out at ceiling level in the apartment below," says Webb. "Another basic test was to put a coin up against the wall and let it go. It fell through the gap as if going into a slot machine."

Asked to carry out a proper survey on behalf of an increasingly concerned Fred Jones, chair of Newham's Housing Committee, Sam Webb and a team of architecture students surveyed more than 50 of the 110 apartments in Ronan Point,

Ronan Point gives a graphic demonstration of progressive collapse. Eyewitnesses described the slab walls and floors of the twenty-two-story apartment building popping out of the building like slices of bread being forced out of a toaster.

including all 22 on the southeast corner that had collapsed. All had serious faults. Worse still, the central staircase and the elevator shaft showed cracks indicating that there had been movement throughout the building. In March 1984, Sam Webb drew up a comprehensive six-foot-long scale plan of the building, plotting the cracks as he had seen them.

"For the first time it became clear what was happening to Ronan Point. In high winds it was beginning to break up. That scenario fitted with the anecdotal evidence of residents telling us about grinding noises and creaking noises when it was windy," recalls Webb. "I deduced that Ronan Point was moving on its lifting bolts and was held up by the 'blast angles' fitted after the public inquiry. The drypack mortar had been crushed—or was never there in the first place."

A building that had undergone the scrutiny of a public inquiry, had been comprehensively repaired and passed as safe by a firm of engineers hired by Newham Council earlier that year, seemed to be proving its critics right: it could not be made safe. The original gas explosion had simply highlighted its overall structural deficiencies. In the end, Ronan Point was just as unsafe without gas appliances in the apartments. Fred Jones was horrified. With the agreement of the tenants he called an emergency meeting of Newham Council's housing committee and forced through a measure to evacuate the building to facilitate a full inspection. With the bottom two floors cleared, he moved council housing officials and engineers into them as an added incentive to get on with the job.

Fire still remained Sam Webb's main concern. When the blast angles were fitted he had pointed out that they were not fire protected. Now, with gaps between walls and ceilings obvious, he doubted anything in the apartments was fireproof. He demanded a fire test. On July 18, 1984, with the local fire brigade on hand and heat and movement monitors set, officials from the Building Research Establishment (BRE) set fire to a sofa in a third-floor apartment. Watching tenants had been told that if there were no problems, the fire would be allowed to burn for twenty minutes.

As television crews and journalists stood with a small group of residents, smoke began billowing from the windows into the summer sunshine. But within twelve minutes the fire crews had turned on their hoses in a desperate effort to extinguish the flames. As Andrew Davenall, Newham's chief engineer, told a reporter from the BBC's *Newsnight* television program: "The deflection reading that we had preset had been exceeded. There was a danger of progressive collapse. We did not want to be responsible."

Nor did anyone else, it seemed. As expected, when the fire reached a temperature of about one thousand degrees Fahrenheit, the concrete slab that formed the ceiling of the apartment started to bow downward as it expanded. This soon exerted lateral pressure on the walls. The bowing was later measured at more than two and a half inches and one of the two precast units forming the slab was found to have split. With uncertain fixings, the wall panels would soon have moved outwards just as the ones in Ivy Hodge's apartment had, sixteen years previously. The effect would have been the same—once one wall panel had gone, progressive collapse was inevitable.

A critical mass of evidence and opinion now seemed to have been reached. Although BRE insisted fireproofing was an option, the tenants were not interested. Another report from the Newham Council's independent consultants, Building Design Partnership, could only have further alarmed them. Even though Ronan Point was barely one-quarter of the way through its intended lifespan, the H-2 joints would need to be reinforced for the second time "if long-term occupation was to be considered." The report added comfortingly that the block "could fail under certain abnormal loads." It did not specify what such loads might be.

By the fall of 1984 the national press had resumed their interest. "Ronan Point lives in a state of uneasy equilibrium, the fact that it is standing does not make it safe," Sam Webb told *The Times.* "Sooner or later something will give. The question to be asked is whether we take these buildings down or gravity does it for us." By the mid-1980s, Sam Webb's complaints fitted into a huge national pattern of failing tower blocks—crumbling concrete, water penetration, panels falling off. Councils were spending millions repairing and refitting them. Indeed, an industry almost as big as that which had built them had now sprung up to repair them.

Tenant militancy had grown accordingly. Organized by tower block, estate, and region, residents were demanding answers beyond structural surveys and repairs. "It seems extraordinary that we are now taking advice from the same experts who gave us all the wrong advice about building these places in the first place," remarked one housing officer in Hackney, London. On October 9, 1984, Newham Council decided to demolish Ronan Point. Sam Webb and the tenants insisted that it be demolished floor by floor so that the joints could be thoroughly examined.

In May 1986 the "infernal tower," as the press had labeled it, was swathed in reinforced polystyrene and demolished as it had been constructed, panel by panel. Sam Webb was employed to advise the tenants. He insisted on Ronan Point being an open site for the press, engineers, architects, housing managers, students—anyone who was interested. For nearly a year the site was like an archaeological dig, every joint photographed and recorded—the excavation of the bones of system-built public housing. "Infinitely better workmanship has gone into the dismantling of Ronan Point than in erecting it," observed Fred Jones wryly.

He could not have known how right he would be. "I knew we were going to find bad workmanship—what surprised me was the sheer scale of it. Not a single joint was correct," recalls Sam Webb. Fixings straps were unattached; levelling nuts were not wound down, causing a significant loading to be transmitted via the bolts; panels were placed on bolts instead of mortar. But the biggest shock of all was the crucial H-2 load-bearing joints between floor and wall panels. Some of the joints had less than 50 percent of the mortar specified.

In its place, quite literally, was rubbish. Cigarette ends, tin cans, bottle tops, bits of wood and newspaper, nails, dust—anything you would associate with a building site. "It was everything they had swept off the floor before putting in whatever drypack mortar they did put in," says Sam Webb, shaking his head. Andrew Davenall, was visibly shocked: "It is much worse than we thought, much worse," he admitted to one news crew at the site.

The failure to install the support bolts properly meant that the wall panels were often resting on the rubbish. It was a message not lost on the demolition workers, some of whose fathers had erected Ronan Point. They told reporters and engineers they felt unsafe. So it seemed did a number of practicing structural engineers. Attracted by the controversial reports emanating from the "open" site, several had come to east London to see for themselves. "When they put their hands in where there should have been solid concrete they just went white because they suddenly realized what had happened," recalls Sam Webb. "For years it had been difficult to get them to believe what we'd been saying...you were asking them to deny everything they believed in. That's a very difficult thing, especially when their careers depended on it."

The obvious questions ranged beyond the initial supervision on the site. If all the joints showed similar problems, why had they not been reported when the southeast corner of the block was rebuilt after the public inquiry? What, moreover, were the implications for other system-built blocks? It soon emerged that even though the government had issued figures for the number of tower blocks that should have had remedial repairs done after the collapse of Ronan Point, it had no list of which blocks had actually had such repairs carried out. "A new list must be drawn up immediately so that surveys can be carried out," Sam Webb told *The Times.* "It's a matter of extreme urgency. People's lives are at risk."

They still are. Though Ronan Point and its sister Larsen-Nielsen-built structures have all been demolished, high-rise system-built apartments remain in use throughout Britain. Recent reports in the architectural press have shown that some such apartments are still connected to gas mains. Moreover, although building regulations specific to progressive collapse have been introduced in Britain, they do not cover structures standing before they were enacted. Some architects and structural engineers are concerned that the building regulations covering progressive collapse remain inadequate and poorly enforced. The Standing Committee on Structural Safety (SCOSS), the body that advises the building regulations division of the British government's Department of the Environment, Transport and the Regions (DETR), remains deeply concerned about progressive collapse. As SCOSS's twelfth report, "Structural Safety 1997–99: Review and Recommendations," noted somewhat laconically: "The long-standing dialogue between SCOSS and the Building Regulations of the DETR on this topic has continued."

TODAY SCOSS IS STILL PARTICULARLY CONCERNED ABOUT MULTI-STORY car parks, assembly halls, shopping complexes, and stadiums which "tend to have a greater vulnerability to progressive collapse." The twelfth report concluded: "Explosion within or external to buildings or other structures is one circumstance in which resistance to disproportionate collapse is essential... structures resistant to disproportionate collapse generally also give some degree of explosion resistance. Robustness, or structural integrity as it is often termed in North America, has become a topic of great importance there since the bomb attacks on the New York World Trade Center and in Oklahoma City."

As the engineers who investigated the Oklahoma City disaster in 1995 (Chapter 6) were the first to admit, few of the lessons of the progressive collapse experienced at Ronan Point twenty-seven years earlier had been absorbed by the regulators and construction industry. That failure would cost many more lives and still threatens millions of others—people who everyday unknowingly live or work in unsafe buildings.

GROUND FORCE

When the Earth Moves

EARTHQUAKES KILL VERY FEW PEOPLE. IT'S THE BUILDINGS THEY DE-stroy or damage that have, over the years, killed millions. Indeed, at no time does a home, office block, hospital, or school—a structure designed to shelter and protect—become more of a potential death-trap than during what structural engineers term a "seismic event." And, despite the development of a whole field of science and engineering designed to predict and combat earthquakes, it could be argued that, historically, it's getting worse. Probably more people have been killed by earthquakes during the past century than ever before.

The reasons are simple. Earthquakes often hit parts of the world with the least resources to predict them and combat their effects. A larger and more urbanized population means bigger and more densely packed buildings and structures— that is, more targets. Hundreds of people were killed in San Francisco in 1906, and hundreds of thousands in China in 1976; buildings felled by earthquakes are the biggest natural-disaster killers of our time. The devastating Turkish earthquake of August 1999, which may have killed as many as forty thousand people and spawned an outcry over appalling construction standards, is but the latest example of a major disaster.

Understanding earthquakes has been the key to resisting them. Ancient beliefs blamed subterranean fires, or wind escaping from caves within the bowels of the Earth. Often these beliefs were linked to the anger of the gods. The Maori explanation, for instance, was simple—Ruaumoko, god of volcanoes and earthquakes, had been accidentally smothered as his mother turned face down while feeding him. He was still spitting and growling. Some groups of Native Americans believed the Earth was supported by a giant tortoise. Logically, earthquakes occurred when the tortoise lumbered forward or stumbled.

Since the beginning of the twentieth century, geologists, then seismologists, have come up with more scientific explanations. About five billion years ago the Earth, a mass of hot gases, began to cool. As it solidified over time, its outermost layer—or crust—cracked. Today this crust can be compared to a cracked eggshell, divided into seven large and twelve small pieces. These are known as tectonic plates which, driven by currents of heat from the Earth's molten core, move continuously, rubbing and pushing against each other. The edges or gaps between these plates are fault lines and it is along these that most earthquakes occur. The friction between the plates causes stress to build up along the boundaries between the two plates. When a stress level sufficient to overcome the frictional resistance of the rough edges of the plates is reached, a sudden slip is felt as an earthquake takes place. The shallower the point of eruption of the slip—that is, the closer to the surface it is—the more dangerous it tends to be.

There are two main bands of seismic activity. The circum-Pacific Belt rings the coast of the Pacific Ocean from New Zealand through southeast Asia and Japan across to Alaska and down the whole Pacific Rim of the Americas—Canada, California, Central America, and through Peru and Chile. The Alphide Belt cuts in from the Atlantic Ocean to travel through the Mediterranean and threaten both southern Europe and North Africa, and crosses Asia from Turkey and Iran to China. This belt has produced some of the most devastating earthquakes of recent times, particularly in China where at least seven hundred thousand lives have been lost in four major quakes. But Italy, Iran, and Turkey have also suffered particularly badly. About 70 percent of earthquakes take place along the circum-Pacific Belt; about 20 percent on the Alphide Belt.

That leaves about 10 percent occurring outside these zones. Such earthquakes can of course be the most destructive, for these occur where there is least local building-code provision for earthquake-proof measures. The frightening reality for structural engineers is that no region of the world is completely free from the threat of earthquakes. Two good examples are a major quake in Charleston, South Carolina, in 1886 on the United States' Atlantic—not Pacific—coast and a series of three earthquakes in New Madrid, Missouri, in 1811–12, about as far as you can get from either coastline in the United States. It is now estimated that these earthquakes would have registered 8.6, 8.4, and 8.7 on the Richter scale, as powerful as any in the last century. A repeat of such an earthquake today—the product of a six-hundred-million-year-old crack in the middle of the apparently solid North American plate—could cause more than fifty billion dollars' worth of damage over a two-hundred-thousand-square-mile area.

Earthquake measurement is a relatively modern science, a mere forty years old. Its emergence has coincided with the development of what is now termed earthquake engineering: the ability of structural engineers to design buildings to resist seismic shocks. This ability is in turn a result of another relatively recent development, engineers' ability to calculate the loads and forces to which individual components of structures will be subjected. Despite all this, the steepest learning curve has been in the "trial-and-error" laboratory that is the real world. Although the failure of many buildings in earthquakes has been completely predictable, this has not always been the case. Many structures deemed earthquake-proof by engineers on the drawing board have proved disastrous failures in the field.

To know how to combat earthquakes, structural engineers need to know what they are combating. They have classified the effects of earthquakes into four categories: ground rupture, ground failure, tsunami, and ground-shaking. It is a measure of their difficulties that of these it is only the effects of the fourth, ground-shaking, that can be reduced significantly by structural engineering itself. Land-use planning based on geological survey is a far more effective weapon in counteracting the other three.

Ground rupture was most clearly demonstrated in the San Francisco earthquake of April 1906. The city, like the State of California, is sliced by the San Andreas fault, a 750-mile-long fissure that is just one of a system of fractures of the Earth's crust between the Pacific and North American tectonic plates. When the forty-second-long earthquake rumbled through the heart of the city, gaps of up to twenty feet opened up in the ground, breaking water and sewage pipes and half swallowing buildings, walls and pavements. Structural engineers quickly concluded that it was impossible to design buildings that could withstand such displacement. The key to avoiding it was knowing where ground rupture might occur.

Ground failure is the result of the shaking of unstable soils. One of the most obvious examples of ground failure occurred in the Alaskan earthquake of March 1964 when an area about one thousand feet wide and two miles long in Turnagain Heights, an elegant suburb on the bluffs facing the Knick Arm Fjord, broke into thousands of soil blocks. Seventy-five homes were destroyed, some of which shook, rattled, and rolled their way fifty feet to the bottom of the cliff. Sandy soils, mud, and infill sites are the most likely candidates for ground failure and it shows itself in landslides, settlements of loosely compacted soils, and liquefaction (which occurs when sandy or moist soils saturated with water momentarily turn to liquid when subjected to shaking). The most obvious examples of liquefaction were in Japan, during the 1964 earthquake in Nigata and thirty-one years later in Kobe (see page 172). Such phenomena often occur some distance away from the epicenter of the quake.

The tsunami is a seismic sea wave, popularly known as a tidal wave, caused by earthquakes at sea. It originates when the level of the seabed drops suddenly, usually along a tectonic boundary. The sea, sucked into the void, creates a succession of waves. With long periods between the crest of successive waves—some can be as much as sixty minutes—relatively small displacements can grow in height several times before they hit the shore. Alaska, Hawaii, and Japan are particularly vulnerable to tsunamis, which have been known to traverse the width of the Atlantic and Pacific oceans and travel miles inland up estuaries and rivers. It was a tsunami that killed 122 of the 131 victims of the 1964 Alaskan earthquake. The same tsunami flooded thirty blocks of Crescent City, California, more than fifteen hundred miles from Anchorage. Early warning and prompt evacuation are the only real defenses against tsunamis.

Ground-shaking, the fourth effect of earthquakes, is the one feature on which structural engineers—as opposed to geologists, seismologists, and planners—can have some influence. The basic rules learned by engineers, mostly as a result of studies in the "living laboratories" of the aftermath of earthquakes, are simple.

Buildings need to be able to absorb the shocks of earthquakes without crumbling in total collapse. But they also need to lose as little of their non-structural materials—concrete panels, parapets, tiles—as possible: for these can be as hazardous to life and limb as the collapse of major structural components. The keys are continuity, ductility, symmetry, and torsion. In layman's terms: how well a building is knitted together by means of joints and frame; how much it can be bent out of shape and remain intact; and how symmetrical it is in its design in order to reduce the concentration of stresses at points that might quickly fail.

During an earthquake a building or structure vibrates in response to the motion of the ground, just like the string of a violin or guitar when plucked. The more rigid a building, the harder it is shaken by the earthquake and the more likely it will be damaged or destroyed. The taller the building, the more it needs to oscillate to absorb the shock if it is to resist the seismic waves. Yet it cannot oscillate too much if it is to be habitable. The whole structure has to be a calculated compromise.

Thus in different earthquakes different buildings will have different amplitudes (or amounts of swing) as well as different periods of vibration (the period being the time taken to return to upright). These differences can be crucial: when swaying, shorter, less flexible buildings can actually collide with taller buildings if there is insufficient space between them, sending debris flying in a phenomenon known as pounding.

To add a further complication to the calculations, the direction of a building's oscillation is unpredictable. It could move in three different planes: forward and backward, from side to side, or, worst of all, it could twist as different materials or features of the design resist the seismic waves in different ways. Earthquake engineers have discovered that a building with a regular plan, something that is symmetrical whether square, circular or triangular, will oscillate without twisting—a potentially life-saving feature.

However, engineers cannot design out resonance, another major threat. When the natural period of vibration of a building coincides with that of the seismic waves of the earthquake, the building is said to be in resonance with the ground. The vibrations from the ground can effectively be amplified each time they are in resonance, magnifying the effect of the earthquake on the building.

The key to all this is the building's ductility—its ability to bend, stretch, and twist without breaking. Steel is the ideal material in earthquake zones because of its ductility. Reinforced concrete, a basic material in so many buildings, is not very ductile and cannot meet most building codes in earthquake zones without large amounts of steel reinforcement. Even when stressed beyond its elasticity in an earthquake, a well-constructed steel-frame structure can survive. It is said by structural engineers to move beyond "elasticity" to "plasticity." At this stage the building will be permanently deformed and damaged but will remain standing.

One of the keys to the reaction of the steel in a building during an earthquake is what engineers call continuity—how well the steel frame is welded together, how well the support columns are connected to the floors or ceilings, how good the joints are. "The energy of the earthquake will simply find the weakest link. The crucial thing is quite literally how well your building hangs together," notes one structural engineer.

There are other solutions beyond the materials used in a building. One is to use dampers in a structure's joints or bracing elements. Dampers act like a car's shock absorbers, soaking up the energy of the quake and dissipating it. Another solution is base isolation: quite literally isolating the building from the ground that shakes it. The theory is nearly a century old, the practice just over a decade—the materials, models, earthquake-motion records, and computer programs making it all possible having taken that long to come together. Base isolation works by fitting buildings with pads made of alternating layers of steel and rubber. Aphrodite is one of the most notable beneficiaries. The Getty Museum in Los Angeles has put her and many other of its most valuable pieces of sculpture on base isolators, allowing them to slide back and forth during an earthquake.

Aphrodite is just one of millions of residents occupying one of the most dangerous earthquake zones in the world—the area around the San Andreas fault. This fissure, hundreds of miles long, is slipping at an average rate of nearly two inches a year. Ever since 5:12 A.M. on April 18, 1906, when a quake measuring 8.3 on the Richter scale hit downtown San Francisco and killed more than five hundred people, Californians have been waiting for the "Big One." Seismology—which includes the all-important science of prediction—has improved phenomenally since the turn of the century but, by the 1970s, the best prediction of a major quake along the fault was still only some time between 1988 and 2018. Moderate tremors in June and August 1988 raised the stakes.

Then, at 5:04 P.M. on October 17, 1989, with much of the San Francisco Bay area focused on the Candlestick Park stadium where two local teams were vying to become World Series baseball champions, time stood still. What became known as the Loma Prieta earthquake measured 7.1 on the Richter scale and lasted twenty seconds. At the time, it felt like the waiting for the "Big One" was over. But the sixty-two thousand fans at the game were lucky. The stands at Candlestick Park had been built to the latest earthquake-proof specifications. There was a collective gasp when the stadium oscillated briefly as the seismic waves traveled through the underlying rock and into the foundation pilings before the structure settled back into its original position.

Others were not so lucky. Bud Minton, a graphic designer, was heading home, traveling south on Interstate 880 along the Cypress Street viaduct section of the Nimitz Freeway, a two-mile double-deck roadway that traverses a poor West Oakland neighborhood. Suddenly his Volkswagen Beetle lurched to the left and Minton found himself fighting the steering wheel. Oh dear, he thought. A flat tire.

It was then that the road started to move. "I could see the freeway ahead rippling toward me in a ribbon-like fashion. Cars were disappearing. They would show up and then disappear," he recalls. Bud Minton had experienced earthquakes before—they were hardly uncommon in the San Francisco Bay area—but nothing like this. He braked, slowed to a stop while bracing himself for an impact from the rear. Somehow he managed to get out of the car. "There were cars on their sides. There were horns honking. There were people screaming and crying. It reminded me of a war zone."

It was clear there were casualties, perhaps scores. The Dodge Ram minivan carrying Cathi Scarpa, a nurse on her regular commute back from work with seven colleagues, had been just seconds ahead of Bud Minton's Volkswagen. They were in the middle of the concrete ripples he had seen coming toward him. Twice the van took off before slamming back into the buckling roadway, then pitching head-on into a concrete barrier and sliding to a halt amid a pile of rubble.

Seconds before, the passengers had been arguing the relative merits of the Oakland Athletics and the San Francisco Giants. Now five of the eight were dead; the three others seriously injured. Among them was Joy Edstrom, a gardener at Golden Gate Park conservatory, who had planted some of the

The Cypress Street viaduct—its columns are splayed out from under the upper deck they were supporting. One observer described it as collapsing like Pluto the dog as he sinks to the ground in a Disney cartoon.

cypress trees lining the street from which the viaduct took its name. "For a moment I actually wondered if I was dead," Cathi Scarpa recollects.

Ron Carter, assistant chief of operations in the Oakland Fire Department, was among the first rescuers on the scene. What confronted him was an almost precision collapse: a mile-and-a-half strip of the upper section of the Cypress Street viaduct had simply pancaked onto the lower section. There was almost no lateral displacement. Carter found the scale of it hard to comprehend. "Such a large structure like that...for such a distance...that was hard to believe." It was even harder to know where to start. The rescue team decided to place a ladder against the first column. "We found a car with four people inside. The two adults were deceased but there were two children trapped inside."

Working conditions for the rescuers in what was now a mere four-foot gap between the two sections of freeway were impossibly cramped. The vehicles and their occupants traveling north had become the filling in a concrete sandwich, the air filled with the smoke and fumes of burning vehicles and fuel. The only design feature that had ensured there was any space at all on the lower deck were the supporting caps. Bridging two supporting columns, these formed frames around the upper and lower freeway at fifty-foot intervals. "They [the caps] were all cement core...extremely heavy. When the top layer came down on the bottom layer the bottom of the cap—the part that hung down about four feet—created a void," recalls Ron Carter.

Crawling around like miners at a coalface, yet sixty feet in the air, Ron Carter and his team braved the aftershocks and the oxygen deficit to try and free the two children in the back of the family car. Using every piece of cutting equipment they had and spurred on by the moans and crying, the crew took three hours to extract the young girl. Her brother was more of a problem. His leg was pinned under one of the caps.

A surgeon quickly made the decision to amputate despite the difficulty of administering an anesthetic. It worked. After nearly six hours Julio Berumen joined his sister in the hospital. "As a firefighter you do whatever is necessary to save a life," says Ron Carter. "Most of the time you can see the beginning and the end of it. There is no hesitation to kick in a door or break out a window in order to save the person. But there were some other challenges associated with this one that made you think."

There were challenges for the investigators as well. Some 42 people were killed and 108 injured in the Cypress Street viaduct collapse. Police estimated that but for the World Series game keeping people indoors, sixty more vehicles would have been on the crucial stretch of freeway. It could have been much worse. Yet, overall, just 62 people had died in the earthquake— a remarkably low number given the time and power of the tremor, a reflection of how well other buildings and infrastructure had withstood the rigors. So why had the Cypress Street viaduct collapsed so completely, so spectacularly, and, above all, so rapidly, giving so few motorists time to get off?

For Steve Whipple, an engineer working for the California Department of Transportation (Caltrans), the investigation started the moment he began crawling around on his hands and knees in the gap between the upper and lower decks trying to make things safe for rescue workers. It was clear that the principal failure had been in the columns that supported the upper deck. Along with the caps these columns formed 124 reinforced concrete frames, 49 of which had failed. The columns, it seemed clear, had failed at the hinges or joints where they met the lower deck of the freeway.

Some of the breaks looked quite clean although reinforcing steel was protruding from many. Indeed, from the air, a pattern was clear. The columns were splayed out from under the collapsed upper deck like the legs of Pluto the dog as he sinks to the ground in a Disney cartoon. The columns had quite literally been forced outwards (see photo on page 165). Meanwhile, most of the columns supporting the bays of the lower portion of the Cypress Street viaduct had remained standing, not only supporting the collapsed upper deck but having withstood the pressure of the force of that collapse when it actually occurred.

Steve Whipple soon saw the probable failure sequence. It had started with the hinges or pins at the bottom of the columns supporting the upper deck. Although these hinges were designed to incorporate some give, they had not been flexible or ductile enough to withstand the forces exerted by the earthquake. "The initial failure started at the bottom of the column. That column was allowed to move outward from the lower deck, hence removing support from the upper structure. Then as the upper structure started to come down, the upper connection at the top of that column also failed."

The hinges or joints in the Cypress Street viaduct had had nothing like the ductility of the stands at Candlestick Park. Steel generally has a high ductility, absorbing the energy oscillations of an earthquake in alternating swings from tension to compression with impressive results—for example, deforma-

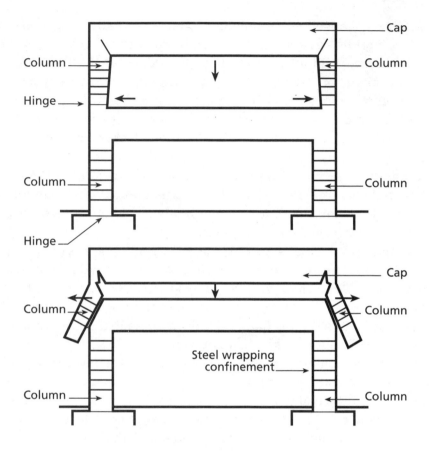

FIGURE 8

The Cypress Street viaduct, before (top) and after the upper freeway collapsed onto the lower freeway (bottom).

tion or damage but no collapse or catastrophe. But at the Cypress Street viaduct the steel joints were very poorly reinforced. As in the Alfred P. Murrah building in Oklahoma (see Chapter 6), there was no continuity: the steel reinforcement bars that would today be required to run through the whole length of the columns stopped at the hinged connection. Moreover, there was little steel wrapping around the columns—what engineers call transverse reinforcement or confinement.

Standing in front of a scale model of a modern bridge column with its steel wrapping, Andrew Whitaker, associate director of the Pacific Earthquake Engineering Research Center in San Francisco, explains confinement by means of a beach analogy. "When you fill a plastic bucket with sand and you turn it upside-down, if you don't take the bucket off the sand castle and you stand on the bucket, the bucket remains essentially intact, the sand remains intact. But if you were to take the plastic bucket off and stand on top of the sand castle it would fail. Transverse reinforcement in this column serves the same role as the plastic bucket. It prevents the column from expanding and collapsing."

There *was* confinement on the columns of the Cypress Street viaduct. Archive film shot during construction in the late 1950s shows the bands of steel being placed around the columns. The trouble was, Andrew Whitaker points out, the bands were too far apart to do their job properly. Quite simply, the twelve- to eighteen-inch gaps between the bands allowed the concrete to bulge out.

In the weakness of the hinges, the inadequacy of the steel reinforcement bars, and the lack of effective confinement, the Cypress Street viaduct was a product of its time. In the thirty years since the viaduct had been built, earthquake engineering had moved on massively. The seismic standards it was built to were effectively obsolete by the late 1970s. "The Cypress Street viaduct was designed using state-of-the-art information but real earthquake engineering only started as a discipline in the early 1960s... prior to that knowledge was more anecdotal than anything else," explains Andrew Whitaker.

The weakness of the columns was a compounded tragedy. Many of those who died were crushed to death under them. Having realized it was an earthquake, they had had time to stop their cars and follow the basic earthquake advice drilled into every Californian schoolchild. Take refuge under maximum support: doorframes, lintels, reinforced columns. As the upper deck of the freeway threatened to fall, logically the maximum support would be around the columns and the supporting caps which made up the frames.

It was actually the worst place to be. As the columns failed, the caps became killers rather than saviors. Some cars had even stopped under these caps, lined up across four lanes of freeway as if on the starting grid before a Grand Prix race, Ron Carter recalls. "When the cap failed it came all the way down to the surface below, crushing all four cars. It came down with the kind of force necessary to crush an engine block paper-thin."

Caltrans knew there was a problem with many of its older bridges and overpasses and had been engaged in a retrofitting program since the San Fernando earthquake in 1971. The only problem was that this was a reactive program that always focused on combating the problems thrown up by the last earthquake rather than the next one. For seventeen years the retrofit program consisted largely of lashing the decks of multi-story bridges together with steel cables to prevent the kind of collapses that had occurred on two freeways in 1971. These cable restrainers, as they were known, were fitted to the Cypress Street viaduct in 1977. However, there was no comprehensive survey of the structure or its foundations at the time and the cable restrainers had no measurable effect on the collapse.

After another earthquake in California in 1987, the emphasis changed and Caltrans began to focus on the vulnerability of columns supporting its elevated structures. An overpass at the junction of two freeways, I-605 and I-5, had come very close to total collapse in 1987 and Caltrans had identified the columns of older elevated structures as the cause. A list of more than four hundred vulnerable structures was drawn up but the Cypress Street viaduct was not on it. A restricted budget meant that priority had to be given to overhead structures supported by single lines of columns. The Cypress Street viaduct was supported by pairs.

But would Caltrans have known just how vulnerable the Cypress Street viaduct was if it had done a comprehensive seismic survey of the double-decked highway? It was a question that official investigators needed to answer. While a long stretch of the northern part of the viaduct had collapsed, large southern sections with exactly the same hinge joints and columns had remained standing. The collapsed section seemed to have been subjected to greater force even though the Cypress Street viaduct was more than sixty miles away from the epicenter of the earthquake. What had happened?

The first clue came from geological maps of the area. The northern section of the Cypress Street viaduct had been built on the soft muddy soils so common in the San Francisco Bay area. Bedrock was some 550 feet down. Further south the geology changed, with the foundations of the support columns of the viaduct here grounded in a dense sand. Ground-motion records and shaking experiments with portions of the viaduct columns left standing after the earthquake demonstrated the difference: the soft soils had amplified the effect of the seismic shocks. The shocks had been up to six times the level the viaduct had been designed to withstand.

"The soft soils beneath the failed Cypress structure act like a bowl of jelly. If you shake the bottom of the bowl of jelly you'll see amplification of the motion through the height of the jelly," says Andrew Whitaker. "Simply put, the motion at the top of the soil—what we call the free field—was significantly greater than the motion at the bottom of the soft soil or the bedrock motion. It was perhaps greater by a factor of between two and four."

But there was something else. The first clues came in the eyewitness statements as to the "giant waves" rolling down the freeway as a result of the vibration In an earthquake a structure can vibrate in three dimensions: up–down, front–back, and sideways. Every building has its own frequency of vibration in all three dimensions, essentially a product of its overall stiffness or ductility. The Cypress Street viaduct had a principal vertical frequency almost identical to the frequency of the earthquake at the collapsed section. It was essentially in resonance with the earthquake; the result: an amplification of the movement and much more forceful shaking. Andrew Whitaker compares it to any weight on a support and a tuning fork. "Pull the top of the weight back and it'll vibrate. When the frequency of the shaking is tuned to the frequency of the tuning fork you see the motion of the tuning fork start to amplify dramatically. That's resonance."

CALIFORNIAN OFFICIALS HAD ALWAYS BELIEVED THEY WERE PREPARED, whether it was the "Big One," a little one, or anything in between. Their counterparts in Japan were even more confident. On a major seismic faultline, Japan had every reason to be at the forefront of earthquake engineering and research during its post-war reconstruction boom. On September 1, 1923, one of the most devastating earthquakes ever—measured at 8.3 on the Richter scale—had hit the Kanto region around Tokyo, killing more than one hundred thousand and destroying more than five hundred thousand homes. As if that was not enough, a huge tsunami or tidal wave had followed, swamping the port of Yokohama. Three days later abnormal twenty-foot waves had slammed into the California coast. It seemed to emphasize the linked precariousness of life on the Pacific Rim earthquake circle.

Japanese engineers visited California after both the Loma Prieta earthquake of 1989 and four years later in February 1994 in the aftermath of what is known as the Northridge earthquake in the San Fernando valley. Studying the effects on design and construction technologies in these "laboratories in the field" had become increasingly popular in the 1980s and the Japanese were particularly keen to analyze the impact of the Californian retrofitting efforts.

The results were fed into their elaborate Tokai Plan, one of the most sophisticated earthquake protection plans on earth. For years, Japanese seismologists had been convinced that their own "Big One" was on its way. They knew where it would be: in the Tokai region in the south of the country between Japan's two biggest cities, Osaka and Tokyo itself. Strain between two active plates, the Philippine tectonic and the Eurasian, had been building up since the last major earthquake in 1854.

The seismologists were right about the strain, wrong about the location. On January 17, 1995, the first anniversary of the Northridge earthquake, residents of Kobe, a port city east of Osaka, were awoken by a terrifying rumble, then a violent shake. "There was a tremendous roar then I felt the tremors," recalls Yumiko Katsuhara, who thought it seemed worse than an earthquake. "I thought Japan had sunk or the planet had been destroyed."

At this point, sinking was a real possibility. The earthquake, which registered 7.2 on the Richter scale, had erupted just twelve miles below the northern tip of Awaji-shima, an island three miles to the southwest in the inland sea facing Kobe. A massive tsunami was a real possibility. In the end there was no tidal wave, but twenty seconds of violent shaking—"It felt like a monster had picked up the house and was shaking it around," recalls Yumiko Katsuhara—did bring just about everything else: gas explosions, fires, and collapsing buildings.

The combined disaster of the Kansai or Great Hanshin earthquake, as it would come to be known, would leave more than 5415 dead; 34,500 injured; and half a million homeless. In material terms it would be the most costly natural disaster ever, with direct losses put at 150 billion U.S. dollars. The severing of water and electricity supplies and the severe damage to road and rail links—the earthquake derailed seven trains and toppled the main elevated highway to the city—in the narrow transport corridor between Osaka and Kobe crucially hindered the rescue efforts. Hundreds died in the fires that followed the earthquakes. The fire brigade was simply unable to reach many of them. When they did the firefighters were often without water.

Pat Underwood is an American college professor who witnessed the earthquake in Kobe. Extraordinarily, she had been living in San Francisco in October 1989 at the time of the Loma Prieta earthquake. "When they recruited me to come here my boss—an old friend of mine—said: 'We don't have earthquakes here like they do in Tokyo, and so we're all going to be safe living here in Kansai.' So it was a shock, a real shock. . . . The damage was overwhelming, it was devastating."

That indeed was the first lesson of the Great Hanshin earthquake: you cannot plan or assess areas on the basis of comparatively low seismic vulnerability in a region of generally high seismic activity. Japanese earthquake prediction and risk assessment had become so defined that the whole country had been assessed and classified. Buildings and structures in each area had been designed to withstand only the force of the earthquake believed possible in that particular region.

The Kansai region had been assessed as being an area of comparatively low seismic vulnerability and of low seismic intensity should an earthquake occur. In other words, bridges, roads, and buildings had not been designed to withstand the sort of earthquake expected in Tokyo or Osaka. In Kansai, there had been much less emphasis on retrofitting of those buildings erected in the 1960s, before the new design criteria developed after the 1971 San Fernando earthquake in California.

Kobe proved that there is no such thing as a designer earthquake or, indeed, seismic design coefficients that could accommodate one. Devastating earthquakes can occur in areas considered to be at moderate or low risk of seismic activity. Probabilities and the lack of any historical record showing that such a powerful earthquake had ever hit the region were irrelevant. "Before the Kobe earthquake our design philosophy was to design structures to resist earthquakes which occur every one or two hundred years," notes Professor Hirokazu Iemura of the Graduate School of Engineering at Kyoto University. "There was a consensus: structural engineers did not design structures against the sort of earthquake motion that occurs every thousand years or so. After Kobe all that has changed."

But something besides the assessment of the risks made Kobe particularly vulnerable. Pressure of space means that much of Japan's coastal land has been reclaimed from the sea. Port Island, Kobe's port and Japan's second largest, had been completed in 1983 by precisely this method. Using mostly decomposed granite soils, the island had literally been built from the seabed of Osaka Bay. Much of the waterfront in Kobe had been extended at the same time.

Port Island and Kobe's waterfront were by far the most devastated areas, with massive land subsidence evident everywhere. With cranes toppled into the water and the quay itself under water, just 8 of 186 berths at the port were operational after the earthquake. This reduced the entire cargo handling capacity of Japan by 10 percent and further hampered relief efforts. The quay wall had moved between three and four meters into the sea, opening a huge gully along the waterfront into which buildings had simply subsided.

Sand boils—eruptions of sand from below the surface—and other tell-tale signs led structural engineers to quickly identify liquefaction, a well-known but only partially understood phenomenon. "Liquefaction is a transient condition where the ground just loses all its strength and stiffness for a few seconds," says Professor Scott Steedman, an earthquake engineer who visited Kobe after the earthquake and continues to research the phenomenon. "The soil grains which are normally jammed together—that's where they get their strength from—they're just temporarily blown apart by the water pressure in the ground. Like a quicksand really. Extraordinarily devastating."

To see the effects of liquefaction you have only to add water to a jar full of sand, put in a weight—say a pebble or stone—and shake the jar. The sand will liquefy in the water and the weight will sink or tilt precariously. The latter was something that Japan had witnessed as recently as 1964, when during an earthquake in Nigata a block of apartments tilted eighty degrees without breaking up. The occupants were eventually able to escape by breaking windows and walking down the face of the building.

This incident had stimulated the interest of a young engineer who by the time of the Kobe earthquake was professor of civil engineering at the Science University of Tokyo. Aware of the problem, he was one of a number of scientists and engineers who has now done extensive laboratory and field tests on the infill materials used in Kobe. "The material used in Kobe consisted of silt, sand, and gravel, whereas our research before the Kobe earthquake was confined to clean sand—uniform small particle sand," explains Professor Kenji Ishihara. "Most of our engineers believed that the material used in Kobe would be stable in the face of liquefaction."

What happened in Kobe set off a further round of research into liquefaction. At the U.S. Army Corps of Engineers research station in Vicksburg, Mississippi, Professor Scott Steedman has been leading the way. The most powerful centrifuge in the world has enabled him to recreate the effects of the Kobe earthquake with the help of a one-meter-long scale model of the city's harbor, complete with quay wall, buildings, sea, and foreshore.

Once the model has been spun to the correct speed, an earthquake is simulated by vigorously shaking the model. What the researchers have learned is that the key to the extent of the damage is controlling the water pressures in the ground. "You can do that two ways: by draining the ground or by compacting the ground. Both measures make the ground more dense," says Professor Steedman. "In that condition it's less likely to develop the level of water pressures that are so damaging."

That conclusion was borne out by what actually happened in Kobe. Land reclamation developments that had used the latest technology in terms of fill compaction and consolidation held up well. Rokko Island in the Bay of Osaka adjacent to Port Island had been completed only two years previously. It suffered some subsidence but generally held up well. Kansai International Airport, built on a man-made island using fill cut from a nearby mountaintop and also recently completed, reported no structural damage. The much older Osaka Airport experienced cracking in a runway and suffered cracked walls. The reality was that the infill of Port Island and the Kobe harbor area had been completed twelve years before the earthquake, using methods that were by then twenty years old.

Using their centrifuge, Scott Steedman and his team are still researching liquefaction. With models of different types of structure, they have built up a huge database chronicling how buildings respond to the phenomenon. For Professor Steedman the motivation is the tragedies of the past—Ronan Point in London, the Alfred P. Murrah building in Oklahoma City, the Cypress Street viaduct in Oakland, Kobe's port and harbor in Japan.

"The issue with all these extreme events, whether they are earthquakes or explosions, is: are you prepared to see the structure collapse? I'm not prepared to accept that," insists

Professor Steedman. "As an engineer, I want structures that protect the people that are inside them. I want to be sure that the building doesn't react disproportionately to the size of the shaking or the size of the blast. It shouldn't topple because that's simply unacceptable."

IV

UNDER PRESSURE

TROUBLED BRIDGE OVER TROUBLED WATERS

Schoharie Creek

"IT'S A LITTLE STREAM TODAY—YOU COULD WALK ACROSS IT WITH sneakers and hardly get your feet wet. Tomorrow, three or four inches of rain and it's a raging torrent. It's a very treacherous creek." George King, a retired volunteer fireman, stands on the bank of what does indeed appear to be a gentle stream. Today, on an afternoon in late spring, the eighty-five-foot-wide Schoharie Creek bubbles and gurgles contentedly as it feeds water down from the Catskill Hills into the Mohawk River. It is a picture of rural tranquillity: this land, like so much of upstate New York, is a world away from the state's famous big city or even the state capital, Albany, a few miles down the road.

The region bears the names and the legends of the first inhabitants of this land: Native Americans who left names like Schoharie itself or that of the nearby town of Utica. The area also bears names that reflect the battles and origins of the early white settlers: Fort Hunter, Amsterdam, Tribes Hill. Indeed, some locals had their own explanation for the creek's menace and rapid changes of mood. One shopkeeper who had been flooded by the angry waters of the creek more than once called it "the devil's piss pot." Others said that Native Americans held it to be a sacred place. When the British took it from them they put a curse on it.

The white man had certainly made something of a habit of interfering with nature in the three hundred years since he arrived in the area. In the early nineteenth century, ox-drawn ploughs and handpicks had dug a 363-mile canal through this area, connecting the Hudson River with Lake Erie by means of eighty-three locks. A stone aqueduct carried the Erie Canal over Schoharie Creek until 1940. The main road, the New York State Thruway—linking New York City and Buffalo, via Albany—also crossed the creek here. The latest means of crossing Schoharie Creek had been the Thruway bridge. Designed and built in the early 1950s, it opened to the ever-increasing flow of traffic in October 1954.

There was nothing remarkable about the bridge. Four-laned, 540 feet long and 112 feet wide, it stood an average height of 80 feet above the moody waters of the creek. With five spans of steel girders supporting a concrete deck on four pairs of piers founded on four concrete plinths, it could have passed for any highway bridge on any state thruway in any part of the country. By 1986, the bridge was carrying an average of 15,519 vehicles a day. The only significant problems with it were now more than thirty years in the past. Shortly after it had opened to traffic, all four plinths on which the piers stood had developed vertical cracks. The segments had been reconnected and reinforced.

In late March and early April 1987 upstate New York had heavy rainfall. At first it didn't seem unusual—spring rains could be heavy here. But this rain did not stop. "The storm just did not move over the Catskill Mountain region," recalls George

King. "Hunter, I believe, is where they measured the peak—eleven inches in twenty-four hours. Probably hadn't experienced rain like that since the hurricane came up through New York State back in the fifties." At that time too there had been serious flooding in the area.

Indeed, floods were not uncommon at all at that time of year in the Schoharie Creek/Mohawk River area. But they were not usually caused by rain. March/April was the time of year when the river was flooded with snow-melt from the Catskills. It was not uncommon to see huge blocks of ice bobbing down the swollen waterways. Sometimes these blocks—three or four feet across—got caught against bridge abutments or rocks, damming up the river, flooding farmland. Now, in April 1987, the rain run-off added to the snow and ice run-off, doubling the risk of serious flooding.

Soon Schoharie Creek was in flood and Montgomery County Sheriff Ron Emery and his helpers were advising people in places like Burtonsville, a few miles south of the Thruway bridge, to evacuate. As he drove around monitoring the flow of the fourteen streams that empty into the creek, the sheriff noticed that there were blocks of ice up to four and a half feet thick in the river. By the end of the week, families in the small community of Lost Valley had four or five feet of water flowing through their homes. Sheriff Emery took to traveling around in an inflatable dinghy powered by an outboard motor.

On April 3, with water lapping over several bridges, the sheriff took one look at the water-flow readings, registering a record flow of sixty-eight thousand cubic feet per second, and ordered the closure of all bridges across Schoharie Creek. These included the Route 5S bridge, Route 161 bridge, and the railway bridge. However, the Thruway bridge was beyond Sheriff Emery's jurisdiction. That was a New York State bridge on a New York State highway. "I notified the state police who work in conjunction with the Thruway Authority—told them I was closing all the bridges and suggested that they inspect the situation with the bridge on the Thruway," recalls the sheriff.

The volunteer firefighters were mobilized at five o'clock the following morning as the county flood disaster program swung into action. Residents had to be roused, told how to switch off their power and secure their heating-oil tanks, and asked if they needed help to leave the village. George King recalls: "We told them that if and when we thought it was time to evacuate, there would be one continuous blast on the siren and they were to get their stuff and go to the designated staging area that was the town hall."

George then left the village to take up a monitoring position by the water's edge higher up the creek. "The sheriff's department usually assigns a deputy to check points along the creek to be able to give us visuals as well as instrument readings. It was nothing to be alarmed about at that particular time. We'd seen it that high before and it falls as quickly as it comes up." Sheriff Emery, mobilizing his emergency troops by radio, agreed. His main concern remained blocks of ice and the public. "People are just drawn to water in such a situa-

One of photographer Sid Brown's dramatic pictures of the collapse of the New York Thruway bridge over Schoharie Creek. "I just held up the camera... kept the button pushed down and got about twenty shots of the collapse."

tion, whether adults or children. We've had tragedies as a result before."

The creek, now something of a spectacle with reports that trees, garden sheds, and even a house had been washed away, was indeed drawing an audience. One interested party was Sid Brown, a photographer for the local paper, the *Schenectady Gazette*. Having heard news of the flooding on television the night before, he had made his way to a good vantage point near the Route 5S bridge. "The water was about two feet below the bridge so I started taking pictures of a couple of firemen standing on the bridge." It was more from a sense of duty than excitement. As yet, Sid felt, it was nothing exceptional.

Within a few seconds it would be. With the rain still lashing down, Sid was bending down at the critical moment to put his camera in his bag to keep it dry. "Suddenly I heard this loud noise coming from the direction of the Thruway. It started like a screaming noise, followed by a great crash. The screaming noise was probably the steel bending and falling," he remembers. He stood up, realizing in an instant that the Thruway bridge was plummeting into the creek (see photo on opposite page). Anything on it was going with it. "The first thing I saw was a truck going in, then two Cadillacs." The raging torrent quickly swallowed them. "I couldn't see the cars, I couldn't see the truck. Everything was gone under the water—it was an instant tragedy."

Upstream, George King was scanning the waters with his binoculars. The fire brigade monitors were following bits of dangerous debris down the creek—a propane tank, a fuel tank of some description, and a small building—worrying about what impact they might have on the 5S bridge which was now under water. "We could see the whole area with the binoculars. I could see a tractor-trailer tumbling off the bridge. My first thought was that the driver was gawking to see what the creek was like and had lost control of his vehicle and crashed through the guardrails." In a split second, George realized it was even worse. "There was a humongous noise, a big splash, and the next thing I knew there was a big gap—two sections of the Thruway bridge had collapsed."

George called the emergency control center with simultaneous priorities: the rescue of those in the creek and the closing of the road to prevent others diving to their deaths. Neither was to be. Although at this point it was only one pier and two spans of the bridge that had been swept eighty feet down the creek, within a minute three more cars had driven into the void. The ten occupants and their vehicles disappeared completely. Rolls of paper, the truck's spilled load, were soon bobbing down the raging torrent. From where George stood, they were the only sign anyone had perished.

Sherry Kline and her family, drawn by the spectacle of the creek in full flood, had arrived to watch it all from an embankment seconds after the bridge collapsed. It took a bit of time for it to register: the bridge was down but the road was not closed. Then they heard traffic approaching the bridge from behind them. Sherry still recoils at the realization of what was about to happen. "Oh my God, they don't know, they don't know that the road is . . . that there's a hole up there, that they could fall in, I thought. We waved our arms, screamed and hollered, and told people to slow down, slow down and stop." The drivers must have sensed the Kline family's urgency. "One truck stopped so fast you could smell his brakes burning," Sherry Kline remembers. "But everybody stopped. Nobody went off that bridge after we started to tell them."

As a truck driver positioned his tractor-trailer across the road to block both lanes of traffic, Sherry Kline's husband went down to the edge of what was left of the bridge. Sherry followed, edging as close as she dared to what was now a precipice. "From the top of the bridge you could see two cars upside-down . . . you could see the tires sticking out of the water. I just remember looking at the massive amounts of water that was going by. It looked like hot chocolate, it was so brown and it was so wide. I remember saying, 'God rest your souls, people, I'm sorry there's nothing we can do for you.'"

By now Sid Brown, the photographer, had decided that another section of the Thruway bridge could collapse at any moment. Trudging through the waterlogged fields on the river bank, he picked a spot about halfway back to the 5S bridge. Certain something was imminent, he did not waste time setting up a tripod. "I just held up the camera and stayed there with my finger on the shutter for about ten minutes. It seemed like about two hours but it was probably only ten minutes," he recalls. "Then it just went. I kept the button pushed down and got about twenty shots of the collapse." Some ninety minutes after the collapse of pier 3, pier 2 and span 2 of the Thruway bridge had gone too. Sid Brown had a sequence of shots that would be crucial evidence in the investigation to follow.

It took divers, police dogs, a big fall in the level of water in Schoharie Creek, and a three-week search to recover the bodies and vehicles of the victims. The roofs of the passenger cars had been flattened to the tops of the seats; one vehicle had been swept nearly a mile downstream. Of the ten victims, one, Edward Meyer of Albany, New York, was not found for a year; another, Evangeline Shive of Manchester, New Hampshire, was swept as far as New Baltimore on the Hudson River. The driver of the truck, John Ninham, of Green Bay, Wisconsin, was recovered from his cab with his seat-belt still attached. He was found two-and-a-half weeks after George King had seen him disappear into the torrent.

THERE WOULD BE NO SHORTAGE OF INVESTIGATIONS INTO WHAT HAD GONE wrong at Schoharie Creek. The New York State Thruway Authority (NYSTA) quickly commissioned two firms of structural engineers, Wiss, Janney, Elstner Associates Inc. of Northbrook, Illinois, and Mueser Rutledge of New York, to do their investigation. The New York State Disaster Preparedness Commission hired Thornton-Tomasetti, consulting engineers based in New York City; while the National Transportation Safety Board (NTSB), the federal government's accident investigation body, began their own investigation under Larry Jackson. There would be no substantive disagreement about the causes of the tragedy at Schoharie Creek, but the number of organizations running investigations did demonstrate something relevant to the causes: the fragmentation of responsibility.

Mike Koob and Peter Demming, structural engineering investigators working for the Thruway Authority, faced several immediate problems when they arrived at the site on the Monday morning. The first was access—there was now no road. "It was all pretty complicated. We had to start by getting permission from the local farmers to build on their land," recalls Mike Koob. Another problem was just the water levels. "When I saw the job site I realized the word 'creek' was a misnomer. The water was about fifteen feet deep—there was a major flood passing through this valley," recalls Peter Demming.

With so much at stake, including the possibility that all bridges in the state and maybe the country would need inspecting and even modification, this would be a complicated investigation. It would start with observations at the site, including the remains and the debris; take in a detailed look at the history of the bridge, its design, construction, and repair and maintenance record; and involve a long study of the collapse sequence, including evidence from photographs and video camera footage. Finally, the investigation would broaden out into analysis of three separate areas: the geology and hydrology of the area; testing of the materials and a model of the bridge; and a detailed analysis of the foundations of the bridge. And all this had to be done in a hurry; New York State had to rebuild a bridge across the creek.

Mike Koob hired a local survey aircraft and mapped the sites of the debris. "An aerial survey gave us a pretty good indication of what happened in what order. It was soon quite clear that the initial loss of support came from pier 3—something that backed up the eyewitness accounts," he says. Sid Brown's photographs were another useful source of intelligence. "The photographer had caught all the major sequences on camera and he had also got a shot of the bridge five minutes before the collapse. Initial inspection seemed to show the bridge leaning."

The investigators' first suspicion was a failure of the bearings. As on all bridges, these had some flexibility to allow the bridge deck to expand and contract with the changes in temperature. On the Thruway bridge each girder on each span had a fixed bearing on the east side and a movable rocker bearing on the west side. But the bearings on the Thruway bridge were unusually high, being about three feet tall, supporting twelve-foot-high girders. This height meant that just a small amount of displacement or distortion could throw the whole structure out of the horizontal plane. Had the bearings popped out?

There was no obvious evidence that they had. Some sixteen of the twenty bridge bearings were recovered from the river, including most crucially all four bearings from pier 3. There was certainly damage. They were bent downward one or two inches over lengths of three to five inches, with the fixed bearings showing more damage than the rockers. But was it cause or effect? The ends of the bearings all showed a similar rounding, suggesting they had all been damaged during the sliding and rotation of the end of the girders during the collapse sequence. There was no obvious individual failure.

Another possibility was metal fatigue. As the water subsided, the investigators collected the debris and, with cranes now in position in neighboring fields, they began to piece together the separate spans of the bridge. All the girders were bent and there were fractures along lines of rivet holes, but there was no sign of metal fatigue or shear marks. Demming and his colleagues became convinced that the damage they saw had been done after the collapse sequence had started. Once again, the finger of suspicion pointed at pier 3 in the middle of the creek. The ends of the third span's main girders, which would have been supported on this pier, were much more bent than the others.

A diver had already taken a look at pier 3—or what was left of it—immediately after the collapse. Needing a better look, investigators decided to dam the creek both upstream and downstream. With the water receding, real excavation and inspection became possible as the underwater architecture of the bridge—the plinths and the wider concrete footing which supported the plinth—became visible. The first thing that

was obvious was the remedial measures taken after the 1955
floods, something the investigators already knew about from
their initial trawl through the repair and maintenance records
for the bridge.

Vertical cracks in the plinths had been fixed by massive steel
staples or reinforcement caps across the top. For more than
thirty years these seemed to have worked, although now investigators did find that one of the staples, drilled and grouted
in, had pulled loose, albeit without snapping. It was pier 3
again. As the water receded further it became obvious why.
There was a huge crack in the plinth of this midstream pier. In
fact, it was a good bit more than a crack. Some five inches wide
at the top, it narrowed but continued right to the base of the
footing. Both plinth and footing of what had been pier 3 were
effectively two parts, the upstream portion resting at an angle
or a lower plane, as if in a hole.

A hole was exactly what it was. As the investigators excavated, they found the upstream portion of the plinth had
dropped about five feet, into a void which reached nine feet in
depth in parts. But the length of this pit was even more awe-
some—some forty feet, stretching right to the middle of the
plinth. In other words, one of the pair of columns that supported the road at pier 3 had been entirely unsupported
because the plinth on which it was based was entirely unsupported. The scale of the erosion at the time of the collapse was
indicated by two trees the investigators extracted from the
hole. Both about twelve inches in diameter, the trees had fallen
into the water after the end of the 1986 growing cycle. Both
were almost certainly debris from the flood.

Erosion of the riverbed had always been a possible cause.
The problem for investigators was getting the evidence. Having excavated, they now turned to the construction and maintenance records. The plans, quite literally architect's huge
blueprints, showed that the bridge had no piles to underpin the
plinths and the footings. Further research produced correspondence between the Department of Works and the contracted
engineering consultancy, Madigan-Hyland, which showed

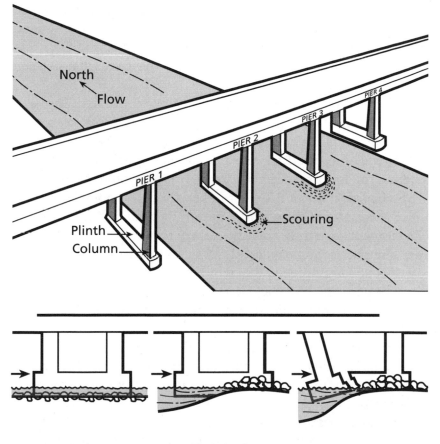

FIGURE 9

Top: The Thruway bridge, showing scouring around piers 2 and 3. *Bottom:* A sequence showing the progression of the scour effect that undermined pier 3.

both parties had considered piles unnecessary. Test boring had shown the riverbed to be fairly firm and resistant, and piles were not an obligatory specification for this type of bridge in the early 1950s.

Without piles, an obvious possible cause of the collapse was what structural engineers call scour. As its name suggests, it's no complicated construction concept: it's simply the wearing away of a material by running water and currents. Anyone can experience the effect by standing still in the shallows at the beach. As a wave recedes, sand behind your heels is eroded and if the current is strong enough they will sink into the holes scoured from underneath them. Exactly the same can happen with the footing of a bridge.

The investigators now faced two obvious tasks. The first was to check the riverbed of Schoharie Creek. Was it as hard and resistant as the contractors had believed? The second was to recreate that riverbed, the flooded creek and the bridge in model form, and see if the Thruway bridge would survive.

The bedrock, about fifty feet below the riverbed of Schoharie Creek, was, as the borings had determined, flat-lying shale and limestone. Above that, and about forty feet thick, was a layer of very compact granular glacial till. It was good foundation material, dense and thick and littered with large boulders, having been compacted by heavy overlying glaciers during the ice age. On top of that there was a thick layer of well-rounded cobbles and sand.

Engineers would have expected the material to show good resistance to scour, but testing was imperative. Three samples of till from different locations in the riverbed were dispatched to Cornell University for tests. All three were composed of different layers or strata and all three remained relatively stable when the water in the river ran at low velocities. But erosion of all three increased dramatically at higher speeds, namely eight to ten feet per second.

The results were somewhat surprising, but did make sense. Not only was pier 3 in the middle of the creek where the flow would have been fastest, but a curve in the river course just upstream from the site of the bridge directed a higher-velocity flow toward it. The trees caught under the concrete plinth of pier 3 could have played a major part, spreading the dimension of the scouring effect into a horseshoe vortex shape around the plinth. Again, it's the same effect you might see standing in the shallows at the seaside: try sticking a piece of wood or a beach spade behind one of your ankles as the wave retreats. Watch how it broadens the scour across its whole length.

Much of this was now borne out by an elaborate model test performed in the Hydraulic Laboratory of the Engineering Research Laboratory at Colorado State University. The three-dimensional, 1:50 scale model recreated a half-mile of river as well as the bridge, complete with riverbed. Using the water-flow readings taken on the day of the collapse, and those taken in 1955 when the bridge's plinths had cracked, the tests showed that the water velocity at pier 3 reached 11.3 feet per second. This was easily enough to cause the rapid scour seen in the tests done by Cornell University.

Yet even then the bridge might have stayed up or, just as crucially, collapsed more gradually. Tests on the concrete showed the pier was well within its design load capability. But a series of test runs with the model quickly demonstrated that when one-third to one-half of the riverbed under pier 3 had been scoured, as investigators discovered it had been at the actual site, rapid collapse followed. The loud, explosive noise heard by some witnesses was the plinth of pier 3 rupturing into two separate pieces. After fracturing, the upstream portion of the broken plinth rotated and dropped into the scour hole.

Lack of structural redundancy in the single-span, two-girder design of the bridge now came into play. There were, quite simply, no alternative load paths or frame action reinforcement provisions built into the design in the event of a loss of support. The bridge spans were non-continuous and were supported by girders that had no other support than the columns of the piers. This, and the tall bearings, guaranteed that the collapse, once initiated, would be rapid.

Ironically, the structural engineers concluded that the collapse was probably made worse by the staples fitted to the plinths after the first cracking problems in 1955. The staples held the plinth of pier 3 together so firmly that they may have prevented the sort of hinge action or gradual slip into the scour holes that had been seen at that time. This meant there would be no early warning, that the first really heavy vehicle crossing the bridge once the scour had reached the critical proportions would bring the whole span and pier down. In the event that vehicle was John Ninham's truck loaded with paper.

Piles driven into the riverbed were one remedy for scour. But there was another. Every civil engineer knows the basics of one of the most common alternatives to piles—rip-rap. It's an unsophisticated solution but no less scientific or effective for that. Rip-rap consists of huge angular rocks, three hundred pounds or more, placed around the base of piers as protection for the plinths and their footings. The next question for the investigators was: had rip-rap been used in the construction of the thruway bridge and, if so, why had it not prevented the scour?

Rip-rap had certainly been built into the design and ordered. Item 80 of the Department of Public Works specification stipulated natural rock with "prismatic shape." Half of this was to be in the form of boulders weighing more than three hundred pounds and the remainder of boulders between one hundred and three hundred pounds. Tests showed that this size rip-rap would have been adequate to resist scour with water flows of no more than forty thousand cubic feet per second. Rip-rap averaging between one thousand and fifteen hundred pounds would have been necessary to be "safe" against the flow velocities experienced in April 1987.

But, more importantly, had the rip-rap even been placed? The design plans were somewhat unclear. They clearly showed rip-rap was to be placed at the upstream and downstream faces of the piers, filling the five-foot holes excavated to position the footing—what engineers termed "a rip-rap backfill to the full excavation depth." But as to the sides of the footings, the drawings were ambiguous. One seemed to show the same protection as for the upstream and downstream faces; another showed a thick surface layer of rip-rap with no infill of the excavated area. "It did not show rock to the bottom, it showed a surface topping. The drawings really didn't show protection all the way to the bottom of the five-foot depth," recalls Peter Demming.

Responsibility for this was equally ambiguous. The NTSB report stated quite clearly that if the bridge designer had wanted rip-rap placed to the bottom of the footing, he had only to specify this on the design plans. The report for the New York State Thruway Authority concluded: "The rip-rap was not placed in the configuration that appears to have been intended by the design, around the sides and on top of the footings." The report went on to add: "The drawings do not convey the importance of the rip-rap" which was "essential for the protection of the footings."

Something else had made matters even worse. When the bridge was built, cofferdams—interlocking boxes made of steel plates—had been placed around the areas in which the plinths and footings were constructed. Designed originally to keep out water during construction, they were also frequently left in place after the completion of a bridge as a sort of container for rip-rap and an additional protection against erosion. At Schoharie Creek they were removed, even though the detail showing the sheeting around the rip-rap was not changed on the "as-built" drawings. Investigators believe they could have made a crucial difference, if only in focusing more attention on the state of the rip-rap during periodic inspections. If inspectors had known the sheeting was not in place, they might have ordered more thorough inspections of the rip-rap.

From the excavations at the site it was difficult to work out exactly how or when the rip-rap had moved from pier 3. However, investigators believed that although some rip-rap may have moved in a series of strong currents or floods before 1987, there had been little if any movement after the collapse, despite the fact that the flood had continued for days. The simple reason was the fall of the debris. This, they were convinced, had acted as a dam around pier 3, a barrier inhibiting any further migration of rip-rap.

What was clear from the excavations was that all the rip-rap found at the upstream end and the sides of the two central piers of the bridge was on the surface. Whatever had been infill around the footings upstream had been torn out. Downstream, as might have been expected, was a different matter. Drops of green paint—presumably deposited while the bridge was being painted—on the top of rip-rap boulders covering the downstream end of the footing at pier 3 indicated one thing. Nothing here had been moved by the 1987 flood.

In all, the investigators managed to recover just 60 cubic yards of rip-rap from in and around the two central pier foundations and the nearby creek channel. The bridge engineer's specifications included the estimate that 700 cubic yards of rip-rap would have been needed to fill the footing excavations and the creekbed at both piers. Indeed, payment records showed that that much rip-rap had been purchased. So, in the case of the two midstream piers, 2 and 3, some 640 cubic yards of rip-rap had been washed downstream—illustrating how inadequate it was in terms of size, weight, and cut.

If it had been placed at all, that is. Peter Demming believes that there was a lack of quality and quantity. "What we did find was a lot of very large river- or creekbed stones, rounded stones, much more easily moved by the river than the angular materials that are typically used for rip-rap. We believe there was a deficiency in the amount of rip-rap and that what was placed was simply not what we would call rip-rap—sharp, angular rock."

Incredibly, given its known importance, it had been suspected for some time that the rip-rap at the Thruway bridge was inadequate. In 1977, Dale Engineering Company of Utica undertook a survey of the bridge for the NYSTA with a view to carrying out a major repair and renewal program. In August, when the water was at its lowest and clearest, they did a series of drop-line (weighted-line) readings around the piers. Pier 2 showed signs of "some minor scouring" and pier 3 had "some scour on the upstream end," the report concluded. Having compared their drop-line readings with the original as-built drawings, Dale's surveyors were convinced that some of the original rip-rap was no longer in place.

Dale Engineering proposed, among a number of other repairs, placing "additional rip-rap around the three piers that are exposed to the creek." A second report, which included preliminary plans and cost estimates, was more emphatic:

"The creek has started to erode the rip-rap protection around the piers," it concluded. "New rip-rap should be installed around three of the piers to prevent further erosion." There is little doubt that Dale's consultants understood the importance of the rip-rap. Indeed, in a partial set of drawings submitted later, they had specified the size of the proposed new rip-rap—six hundred pounds, double the median size of that originally placed.

But not everyone else understood its importance. In 1980, an NYSTA employee who was not an engineer was given the job of finalizing the rehabilitation plans for the bridge. The employee later told National Transportation Safety Board officials that he had visited the bridge before taking any decisions. The water was so low that he could walk around piers 1, 3, and 4. Seeing no scour holes or depressions, he returned to his office in the NYSTA's Bureau of Construction and Design and directed a draughtsman to remove all the references to new rip-rap on the plans. It was also removed from the costing estimate. No one else was consulted; no one raised any objection.

The rehabilitation work was completed by July 1982 with no new rip-rap added. The only significant maintenance work carried out subsequently before the collapse was the painting of the steel superstructure—with the green paint found on the downstream rip-rap. Inspections in 1983 and 1986 failed to pick anything up; the bridge was given an overall rating of 5 on New York State's Inspection User's Manual scale. That meant: "Minor deterioration and is functioning as originally designed."

Yet still the chapter of misdiagnoses, miscalculations, and mistimings continued. In 1986 the assistant superintendent of Thruway Maintenance for New York questioned the specific ratings given to the piers. He made a personal visit with the assistant division engineer but neither noted any problems with scour or rip-rap. Incredibly, by the mid-1980s the bridge had not had a single underwater inspection since construction. Until 1985 it did not qualify for one—Schoharie Creek was not considered deep water.

An underwater inspection was finally scheduled for 1986, then postponed until mid-1987. It was due a few weeks after the bridge collapsed. For the ten people who fell to their deaths in Schoharie Creek, that was a few weeks too late.

A SUCCESSFUL
FAILURE

The Vaiont Dam

STORING WATER BEHIND EARTH AND ROCK EMBANKMENTS FOR DO-
mestic or agricultural purposes is one of man's oldest
uses of engineered structures. It's also one of the most dangerous. Damming rivers or lakes sets structural engineers and
architects in direct opposition to nature in a way no other field
of the construction business does. Dam failure has over the
centuries had many causes, from design miscalculations to
damage by rats. The results can be devastatingly similar.
Dams tend to be in valleys: the populations who use the water
or farm the land tend to live close by. Dam failure is many people's worst nightmare, often reinforced by films and news reports of such disasters. Being swept away by a torrent of
water is powerlessness personified.

The keys to the safety of dams are design, materials, and
maintenance, the same as any other structure. But the compli-
cations are greater. Common problems are piping—the growth
of tubular, leaking cavities beneath the foundations; overtop-
ping, when the water level in the dam's reservoir rises above

the crest of the dam; and foundation failure due to uneven settlement, concrete deterioration, insufficient strength, or erosion. A common problem in earth-filled dams is the failure to compact infill material sufficiently—a feature unlikely to be detected by means of random testing, which is inevitably confined to small, specific areas. In the words of one famous dam engineer, Karl Terzaghi, "The design of a dam is not finished until the construction is complete."

It was a dam failure that caused one of the worst-ever civil disasters in the United States. In May 1889, a forty-foot wall of water traveling at more than twenty miles per hour destroyed most of the town of Johnstown and six villages in the Allegheny Mountains of central Pennsylvania. More than twenty-two hundred of the town's thirty thousand inhabitants were drowned or burnt to death, debris piled against a railway bridge below the town having caught fire. The cause was the South Fork Dam across the Conemaugh River, an earth and rock dam completed in 1853, which was some 931 feet long and 72 feet high. Originally it had supplied water to the Pennsylvania Canal, but had not been used for years when it was sold to the South Fork Hunting and Fishing Club of Pittsburgh in 1880. The new owner, Benjamin Ruff, intended to make what was still one of the largest man-made lakes in the United States the centerpiece of a private hunting and fishing club.

The original design of South Fork was competent and careful. Two-foot-thick layers of clay and earth had been laid, each being allowed to sit under water for several days to fuse into a watertight barrier. Water flowed into the canal through five cast-iron pipes set in an arched stone culvert, and a spillway some eighty-five feet wide was cut through the rock near the eastern end of the dam as an overflow during periods of heavy rain. But poor maintenance led to a partial collapse of the stone culvert running under the dam in 1862.

Within a couple of years of purchasing the dam the new owners had made major modifications. They removed the outlet pipes, lowered the dam by two feet to widen the roadway that traversed it and constructed a trestle bridge across the spillway to prevent loss of fish. The alterations were to prove disastrous. The capacity of the spillway was reduced to barely one-third of its original and the material used in a series of "repairs" was laughable—the uncompacted fill was later discovered to consist of mud, tree stumps, brush, and even horse manure.

On May 28, 1889, the worst storm ever recorded hit the Johnstown area, with about eight inches of rain falling in just twenty-four hours. On the night of May 30, the lake rose two feet. Eating breakfast the next morning, John G. Parke Jr., the club's resident engineer, watched the water continue rising to the top of the dam, despite efforts by his employees to create a new channel. By 11:30 A.M. water was pouring over the dam. Parke, knowing the likely outcome, rode his horse to a nearby village and sent an urgent telegram to officials in Johnstown.

The dam might collapse, he warned. No one paid the slightest attention. After lunch Parke returned to find a hole in the dam that was widening rapidly. "Before long there was a torrent of water rushing through the breach carrying everything before it. Trees growing on the outer face of the dam were carried away like straws," Parke recalled. At 3:10 P.M. a four-hundred-foot-wide segment of the dam crumbled. Many of the residents of Johnstown and its surrounding villages were doomed.

THE JOHNSTOWN DISASTER ILLUSTRATED A BASIC PROBLEM WITH DAMS IN the United States: large numbers were privately owned, unregulated and uninspected. Many dams are by-products of industries such as mining or agriculture, industries that are often based in remote areas. A dam at Buffalo Creek in West Virginia was one such dam. In 1972, the dam, which enclosed waste water from a coal mine, collapsed, killing 125 people. In 1980, the Federal Emergency Management Agency (FEMA) completed a huge survey and concluded that the majority of nonfederal and privately owned dams in the United States were unsafe. They were "poorly engineered, badly constructed, and improperly maintained."

Ironically, it was the failure of a federal structure that led to the most major dam disaster in the United States. On June 5,

1976, just nine months after the reservoir began filling, the Teton Dam in Idaho broke and a fifteen-foot torrent of water flooded three hundred square miles of the state, swamping the communities of Newdale, Rexburg, and Teton itself. Although the towns had been evacuated, eleven people died, more than two thousand were injured, and some twenty-five thousand were left homeless as part of a property damage bill put at over a billion dollars. Like so many catastrophic dam failures, there was little evidence left. Both the construction itself and the immediate environment had been swept away.

What was known was that seepage from the dam had been taking place for two days before the collapse. On the day itself, a muddy trickle had appeared on the face of the right downstream embankment. The trickle had grown to a steady flow within hours as the piping widened to a tunnel. What was also known was that there had been many small cracks in the dam that had been ignored—the design specifications stated that a tight seal could be maintained by ignoring cracks of less than twelve millimeters in width. Furthermore, at the time of the collapse the outlet works needed to maintain a moderate reservoir fill-rate were still incomplete even though the reservoir had been filling for nine months.

Two independent investigations and a panel of ten engineers appointed jointly by the federal and state governments concluded that the principal cause of the disaster was embankment design at an inappropriate site. It was well known that the Teton River site was difficult geologically. It was made disastrous by certain design features of the 3050-foot embankment. The ten million cubic yards of earth and rock-fill in the embankment were not adequately protected from erosive seepage, nor was there adequate provision for drainage. The designers went to great lengths to keep water from seeping through the dam but did almost nothing to "render harmless whatever water did pass." Piping had eroded the base of the embankment's impermeable core material in the keyway that was cut into the right abutment (the part immediately adjacent to the canyon walls). Water then burst through the dam's downstream face because it was inadequately drained, the investigators theorized.

However, the specific cause of the piping could never be determined with absolute certainty, the failed section having been destroyed in the flood. As such, the Teton Dam disaster illustrated the real difficulties of forensic engineering in regard to dams. Walking the site, complicated computer analyses, detailed geological surveys—sometimes nothing provides a definitive answer.

Nine years after the disaster, in August 1985, dozens of world-class engineers gathered at Purdue University, Indiana, for an International Workshop on Dam Failures. Disagreement on what had happened at Teton was as sharp as ever, with participants able to agree on little more than seven possible causes. Reviewing the conference, the trade journal *Engineering News-Record* concluded: "Even without the raised voices it was clear that when structures are made of uneven natural materials, founded on natural materials, and acted on by natural forces, their behavior will always be a matter of speculation."

The Malpasset Dam disaster in southern France was not dissimilar. Designed in 1951 by Frenchman André Coyne, considered one of the most innovative dam designers of the age, the Malpasset Dam stretched across a canyon of the Reyran valley in the French Riviera. It was the thinnest arch dam in the world. Unlike gravity dams, which rely on bulk and foundation to support their weight, arch dams achieve much of their strength and stability through their curved shape.

The key is the quality and condition of the rock that forms the dam's abutments. In the location in the Reyran valley chosen by the geologists, the rock seemed ideal. By 1954 the dam, 181 feet high and with 639 feet of elegant curve between the two abutments, was complete. Malpasset was less than 5 feet thick at its crest and less than five times thicker at its base, with different angles of curve on its upstream and downstream faces. On its right bank the dam abutted a rock face; on its left it sported a concrete wing wall.

The disaster followed a similar pattern to that at Teton in Idaho. After five days of heavy rain, some time after 9 P.M. on December 2, 1959, the arch cracked. It was described as looking somewhat like the blunt end of an egg hit by a teaspoon. There was no warning, no tell-tale sign, just the deafening roar of sixty-five million cubic yards of water squeezing through the narrow Reyran valley. The force of the water carried the three-hundred-ton blocks of concrete that made up the dam up to a mile downstream. Roads, railway bridges, and a four-lane highway were all washed away before the torrent hit Frejus, a town five miles downstream. Some 421 of its inhabitants drowned.

The disaster puzzled the experts. There was little to go on except that the left bank was totally destroyed, the wing wall demolished. This, and eye-witness reports that the dam had started to break up near the center of its curve, raised an obvious question: had the left-side abutment shifted, causing tension and cracks in the concrete blocks that made up the curve?

The design concept and strength calculations of the dam were checked, then checked again. The joints between the concrete blocks were tested and found to be as strong as the blocks themselves; the bedrock to which the concrete blocks were attached was sound. Indeed, blocks were still attached. In the absence of any rational explanation, the French public began to latch on to more improbable scenarios: an earthquake, an explosion, sabotage, or even a meteorite.

A consensus began to emerge that the failure was due to movement in the abutments of the dam. But why or how? The whole episode descended into a morass of recrimination and legal charges about responsibility. A panel of experts' report was rejected; a second panel reported four years after the accident and Jacques Dargeou, the engineer who had accepted the dam on behalf of the French Ministry of Agriculture, was acquitted of charges of involuntary homicide in 1964. Despite the lack of definitive evidence, relatives of the victims then brought charges against four engineers, including Jean Bellier, a partner and son-in-law of André Coyne. All four were acquitted. Coyne himself had died six months after the Malpasset Dam failed, shattered by the whole tragedy.

It was only as a result of the protracted court cases that something like a complete picture began to emerge. For some time the focus had been on the rock to which the abutments were attached. The most likely culprit seemed to be a thin seam of clay, just one and a half inches thick, in the rock adjacent to the left bank of the dam. This, it was surmised, acted as a lubricant, causing the foundations, then the arch of the dam to shift. Vibration may have exacerbated this movement. A security guard at the dam said he had felt powerful shock waves caused by blasting for a highway nearby just days before the disaster. The detonations, it turned out, had been taking place as little as 260 feet from the dam, using as much as two and a half tons of dynamite at a time.

So was anyone responsible? Malpasset demonstrated once again the crucial importance of geological surveys in dam building. All dams, and the water they retain, can alter the stresses in the surrounding rock, but this is particularly true of arch dams. Geological soundings are entrusted to a geotechnical engineer whose basic method of working is to analyze rock samples taken from borings drilled into the rock. Such borings had failed to discover the seam of clay. Many experts concluded that the pre-construction boring program at Malpasset had been nowhere near thorough enough.

Max Jacobson, head of the first investigating panel, blamed André Coyne himself, believing he had cut corners. Coyne, he said, "had been misled by his own genius." But the court cases revealed that later warning signs had been ignored too. Four weeks before the disaster, a routine photographic survey of the dam had revealed changes in the shape of the arch in twenty-eight separate locations. The Ministry of Agriculture's engineer had not considered the revelation urgent but had nonetheless written to André Coyne suggesting an inspection of the whole structure. The letter was in the mail on December 2 when the arch shattered.

AS THE MALPASSET DAM WAS COLLAPSING, ANOTHER ARCH DAM WAS nearing completion on the other side of the Alps. High up in the Vaiont Gorge in the Piave valley, a hundred miles north of Venice, the Vaiont Dam would supersede Malpasset as the thinnest high-arch dam in the world. More poignantly, Vaiont would be the center of a bigger disaster than that at Malpasset. In October 1963, it would in fact be the cause of the biggest dam disaster in history. Like Malpasset, it would be the result of the failure of geotechnical engineers to fully understand the valley in which the dam had been built. But there would be one important difference: the Vaiont Dam would not fail. In fact the disaster would simply demonstrate its strength and resilience. This would be a construction disaster without a structural failure—a unique occurrence.

Unlike the Malpasset tragedy, there could not have been more warning of impending disaster in the Vaiont Gorge. The alarm was first raised in the autumn of 1960, just as construction was being completed. In the afternoon of November 4, following a week of heavy rainfall, there was a massive landslide from Mount Toc, which formed the left bank of the dam reservoir. "My parents' house faced the slope of the slide. We had just come back from school and were eating polenta and dry salted cod," recalls Gervasia Mazzucco. "I'll never forget it. A part of Mount Toc just fell away, actually the most beautiful part . . . it happened very suddenly."

An estimated one million cubic yards of rock ended up just 330 yards from the dam itself, effectively cutting the capacity of the reservoir in half. It was not completely unexpected. Indeed, the designer of the dam, Carlo Semenza, had had growing reservations about both the location and the scale of the project almost from its inception. In June 1957, just a month before construction had started, the local electricity company building the dam, the Società Adriatica di Elettricità, Venezia (SADE), decided to raise the proposed height of the dam by some 30 percent. The increase would almost treble the capacity of the dam and raise the hydroelectric-generating capacity of the structure accordingly.

Carlo Semenza was nervous. He ran the proposed changes of plan past Professor Giorgio Dal Piaz, who had done some geological feasibility studies looking at the limestone walls of the deep, narrow river gorge nearly thirty years earlier. Semenza had been worried about the deep fissures he had found and the possibility of landslides. Dal Piaz was unequivocal. "I have to confess the proposed changes make my blood run cold. I have tried to write a study for Vaiont but the facts aren't to your liking."

Carlo Semenza agreed to the proposed changes but Dal Piaz's comments just seemed to have heightened his concern. In August 1957, an Austrian expert, Leopold Müller, credited with developing a new science known as geomechanics or rock movement, was called in to advise on how the future stability of the reservoir banks should be determined. After a quick inspection of the geology of the surrounding mountainside, Müller reported that it was possible the reservoir would cause slides, some of which might amount to "as much as one million cubic meters in some parts of the future banks."

Müller's opinion could only have intensified Carlo Semenza's worries. But there were other views. In a return to the valley in October 1958 to examine the stability of the left bank in relation to a new road being built there, Dal Piaz concluded that the rock in the area was fractured but it was in place. Local detachments were to be expected but these would not be of "a serious magnitude." And an independent team appointed by the Ministry of Public Works in Rome had given the work a clean bill of health earlier that year. The team had included a geologist, Professor Penta, who worked for SADE.

Then something happened that Carlo Semenza felt he could not ignore. On March 22, 1959, there was a major landslide at the nearby Pontesei reservoir. One man was killed when nearly four million cubic yards of rock slid down the mountain as maintenance men were emptying the reservoir. Could the increase or decrease in the pressure of the water in such valley reservoirs cause landslides? Would the increased capacity he had agreed to at Vaiont be a major threat? The construction of the dam was nearly complete. It was only a matter of months before the Vaiont reservoir would be filling up.

Semenza recalled Müller. In June 1959, Müller recommended a detailed survey to assess the stability of the Vaiont reservoir's banks and outlined how it should be carried out. Realizing SADE's official geologists had vested interests, Semenza turned to his son, Edoardo. Since graduating in geology five years earlier, Edoardo had done several small surveying jobs for SADE. Having recruited a friend, Franco Giudici, Edoardo walked up and down the steep banks of the soon-to-be-flooded valley.

There were plenty of indications of previous landslides but one mass on the left-hand bank attracted his particular attention. It had strange striations on the surface and huge fissures into which water flowing or falling above it disappeared immediately. Edoardo identified the area as what he termed "an uncemented mylonitic zone" extending about one and a half kilometers along the bank of the Vaiont Gorge. On the right-hand bank there was another smaller mass that showed the same characteristics.

Today, Edoardo Semenza is in his sixties and lives in retirement in an elegant house in Ferrera, near Bologna. After a distinguished thirty-year career in academia as a geologist, he has had many years to reflect on what he believes his findings meant. "This was the site of an ancient landslide which had obstructed the valley, creating a lake which had expanded toward where the town of Erto is today. Where the landslide had occurred a small valley and river had evolved," he says, leafing through aerial photographs of the site. It was, Semenza concluded, a classical example of an epigenetic riverbed, literally a riverbed that had originated from something, in this case a landslide. "I surveyed this area down to a depth of ten meters and found large deposits of river gravel in layers—typical of river sedimentation."

What Semenza did not know was how the rock mass he had focused on was being supported. "I just thought it cleaved to the mountain horizontally....I thought if the gradient was steep enough, water could start the movement again." In other words, if there had been a major landslide there before—big enough to block what was today the valley—the right conditions could start another one. The question was: if Semenza Jr. was correct, what were the right conditions?

Edoardo remembers his father being shaken by his findings but also puzzled. Who was he to believe, his son, whom he trusted implicitly, or Dal Piaz and Professor Penta, the geologists who worked for his employers, SADE, both of whom quickly dismissed Edoardo's findings? Demonstrating his natural caution, Carlo Semenza opted for further research. "He didn't want to be blamed for preventing the completion of the dam without definitive evidence," Edoardo Semenza sympathizes. "It didn't seem necessary to stop the construction. Prevention was the best solution."

As work finished on the dam, another geologist, a Professor Calois, produced another opinion. He refuted Semenza Jr.'s arguments, claiming his survey showed that the lefthand valley wall consisted of "extraordinarily firm *in situ* rock—covered with only ten to twenty meters of loose slide material." Thus the hypothesis of an ancient, very deep slide of the *in situ* rock became very improbable. Edoardo, by now doing test borings to depths of up to 187 yards—as deep as he could go—objected in a letter to his father. For all intents and purposes it was too late: by March the reservoir was full to a depth of 651 yards. Ominously, as it was filling, small rock-falls took place in two separate locations.

It had been agreed to install monitors to track any movement in Mount Toc, and now, as the first of these were being installed in June 1960, Semenza Jr. and Franco Giudici presented their formal report. With the borings they had much more evidence, but their conclusions remained essentially the same. The mylonitic zone at about 2050 feet on Mount Toc was evidence of an ancient slide. They argued that the whole mass below Pian de Toc, between two points known as Casera Pierin and Colomber, could slide. Even more alarmingly, Semenza and Giudici argued that this movement could be induced by filling the reservoir. That was exactly what was happening as they wrote. Having reached the first test height of 650 yards, SADE engineers had moved on to the second: 711 yards.

Despite another report from Dal Piaz reiterating his earlier findings—no evidence of movement in the rock on the left bank—Edoardo Semenza stuck to his guns. Indeed, he was writing up further studies defining the boundaries of the old slide mass when it happened. On November 4, 1960, about one million cubic yards of rock and shale slid into the reservoir from Mount Toc after a week of heavy rain. That slide, the one witnessed by the then nine-year-old Gervasia Mazzucco from her parents' window, brought another ominous warning. The slide mass caused a seven-foot-high wave as it displaced the water.

The slide was accompanied by what geologists call "creep"—slight, gradual movement—over a much larger area. Inspection revealed a whole pattern of cracks upslope from the top extremity of the slide. They stretched eastwards and seemed to form the shape of a letter "M" about a mile wide. As geologists and dam technicians met, they agreed that the fractures marked the real extent of the slide and another one was probable. Indeed, the M was getting bigger, as Tito Speranza could testify. He and his brother ran a small gravel excavation firm near the dam, and he watched the outline with its big dip in the middle and almost symmetrical humps on either side. "It was obvious we were heading for a catastrophe. Even with the naked eye you could see the mountain was moving."

The only question now was how to stop it. Leopold Müller, as a leading expert on preventing landslides, was called in again. It was impossible to stop the slide, he argued in a report completed in February 1961. Control in terms of both speed and volume was the only option. His main recommendations were a controlled and slow lowering of the reservoir level and the construction of two drainage tunnels underneath the sliding mass. Water was the key to provoking movement as the rain and the filling of the reservoir had proved; getting rid of it or reducing its effect was the only option. Edoardo Semenza agreed. It was exactly what he had argued.

SADE did embark on a series of remedial measures and tests. The first was to start work on a bypass tunnel, sixteen feet wide and more than a mile long, to ensure water could reach the outlet works of the dam in case of future slides. SADE then installed a grid of sensors on concrete pillars extending two and a half miles upstream to warn of any future movement. Meanwhile, the potential slide area was explored by means of drill holes to a depth of some three-hundred-plus feet and two drainage tunnels. Neither measure detected any sign of a major slide plane. For added reassurance, Professor Augusto Ghetti was put in charge of tests on a scale model in Nove, just twenty miles from Longarone. Finally, and most crucially, the reservoir level was lowered, from 711 yards to 656. The "creep" slowed.

Carlo Semenza was not convinced. In April 1961 he wrote to a former university teacher. "I can't pretend that the problems of the slide haven't been worrying me for months. Things are out of our control and there are no adequate practical measures. After many successes I am now faced with something that is so big it is out of my hands." In a little over six months Carlo Semenza was dead. Much of his natural caution and concern went with him. His son Edoardo had neither his status nor influence. "While he was in charge things were done in a certain way," notes Edoardo ruefully. *Après moi, le déluge.*

With so little movement being recorded, the left bank of the Vaiont reservoir seemed safe again. In the next two years, as SADE engineers became convinced there was no real danger, the reservoir's level gradually rose. In December 1961, permission was granted to raise the water level to 694 yards—the level at which the cracks and slides had started to appear in October 1960. Movement was negligible. From February 1962 the water level was permitted to rise again to 711 yards, then in June to 733 yards, and finally, in October 1962, to 760 yards. By now the speed of movement as measured by the cracks on the left bank was about four-tenths of an inch a day. In November, the reservoir rose to 765 yards in very heavy rain. The effect was minimal: maximum movement of just under half an inch a day.

Throughout the winter of 1962–63 the reservoir was raised and lowered and movements monitored. Müller seemed to be right: the lower the water level, the less the movement on the bank. The engineers from ENEL—the Italian electricity monopoly which had taken over from SADE when all power-generating capacity was nationalized in March 1963—seemed to have acquired the key to the safety of the Vaiont Dam. If there was a real danger of movement the solution was simple: lower the level of the reservoir. Meanwhile, the political and economic imperatives of power-generation in a now booming economy remained foremost. Nationalized, Vaiont had to pull its weight for the national grid.

The engineers were reassured by the results of Professor Ghetti's experiments on his scale model. "The reservoir is relatively safe up to 700 meters [765 yards] above sea level. That is the absolute safety limit but we must do more tests to see the effects of the wave when it goes over," he concluded. "If the slide happens at 715 meters [782 yards] the wave would be huge and villages inside the canyon would suffer." The terminology seemed ominous: why "when it goes over" rather than "if"? And why, within months of Ghetti's warning, did the Ministry of Public Works in Rome give ENEL's engineers permission to raise the water level to 782 yards?

By late July 1963, there were real worries among the local population, if not the engineers. The water in the reservoir seemed strangely muddy and Mount Toc seemed to be emitting an odd subterranean roar. Local officials reported public concerns to ENEL staff in Venice and provincial government representatives in nearby Udine. Before they had any reply earth tremors were being felt. In the first week in September, four individual markers on the left bank moved between twenty-two and a half and twenty-seven and a half inches in just one day. Yet that same week unseasonally heavy rain raised the reservoir to one of its highest-ever levels—787 yards.

Trees were moving, the mountain was moving, livestock, sensing danger on the northern slopes of Mount Toc, were moving—everyone, everything was moving except those in most danger: those living in the valley below the Vaiont Dam. Another week of heavy rains meant that by October 8 the mountain was slipping at a rate of about eight inches a day. By now the reservoir was finally being lowered, but only at a rate of about three feet a day. To many it seemed far too little, far too late. Run-off from the heavy rains made the move almost pointless. The water level in the reservoir was scarcely lowering and the mountain did not stop moving.

At 10:41 P.M. on October 9 there was a loud rumbling, the earth shook violently and then a howling wind broke across the gorge. Some 353 million cubic yards of rock rumbled *en masse* at a speed of up to 82 feet per second into a water level of 766 yards. It filled the entire Vaiont reservoir in fifteen to thirty seconds.

The consequences were obvious: just what Professor Ghetti had foreseen. The first tidal wave—measuring an estimated 328 feet from trough to crest—cleared the dam and crashed down into the narrow gorge below. The gorge acted as little more than a funnel: the pressure forcing the water out was even greater than that of the atomic bomb at Hiroshima. The air-blast, accompanying decompression, and what was now a 230-foot wall of water hit the town of Longarone head-on within three minutes of clearing the dam.

Tito Speranza had been delivering gravel to the dam all day and passed through Longarone some minutes after 10 P.M. on his way home to the tiny hamlet of Roggia. He had called in at a bar opposite the station and started to read the newspaper. The bar owner told him to take it. That saved Tito having to go into Longarone to buy a copy as he had intended. It was a gesture that probably saved his life.

Shortly after he arrived home there was a series of roars. "Longarone was all lit up. I still shiver, recollecting that image in my mind. It seemed we could touch the roofs of the houses with our fingers," Tito recollects. "Then we saw a cloud rising from the village, heard another inexplicable roar then a sudden terrible silence. I turned toward Longarone but it didn't exist any more. There was nothing left—only the moon in the sky."

Tito Speranza drove his van to a crossroads two hundred yards from his house. "Dante's *Inferno* was nothing compared to that view. A boy came toward me and said: "Tito, everybody's dead!" I realized then that my wife had lost all her family."

Gioacchino Bratti, a university student, had been studying for a French literature exam in a small village two hundred meters up the mountainside from Longarone. He heard the roar, realized it was the dam and ran with his mother up the narrow, cobbled street. "Although there was no moonlight we could still distinguish the huge mass of water in the dark. In fact we were getting wet, the water was splashing everywhere. By the time we reached the top of the village everything was over."

Longarone's population of four thousand had been considerably swelled that evening, the bars packed with men watching the European Cup match between Glasgow Rangers and Real Madrid. By 10:45 P.M. nothing was left standing. As the survivors became the first rescuers, the scale of the disaster became clear. Gioacchino Bratti came down the mountain, having collected a pickaxe and a torch to make his way over the debris. "We walked through the ruins and reached the town hall, the center of Longarone. I looked around but there was literally nothing left. Longarone was destroyed."

The would-be rescuers found few survivors to actually help. As dawn broke it became clear that the tall natural-stone houses, the offices, bars, trees, roads, the church and its steeple, the railway line, even the twisting streets of what had been a quaint Alpine town, had been swept away or buried in a macabre desert of rock and slimy mud. Corpses hung in the branches of uprooted trees, limbs protruded from the mud. "The dateline of this story from the valley of death should be Longarone. But Longarone is a town that died...in twenty minutes of complete and absolute destruction," wrote Brian Park and Roland Flamini for the British *Daily Express*.

It soon became clear that other villages and hamlets in the Vaiont valley had fared little better. Villanova, Codissago, and Pirago were all badly hit as were the communities on the right-hand bank of the reservoir. Here Frasegn was totally de-

stroyed by a secondary wave that ran 650 feet up the bank on
its way to Casso and Erto, more than 820 feet above the level
of the reservoir.

Gervasia Mazzucco, the schoolgirl who had witnessed the
original slide in 1960, still lived in her parents' three-story
house, one street up from the bottom of the village. Awoken by
a terrifying roar, she was petrified but was quickly pushed out
into the street by her brother and grandmother, both screaming
at her to run and save herself. "I remember my grandmother
shouting: 'It's the end of the world, I am old, you are young, run
and save yourself. I will wait for death.' I'll always remember
my anguish. It's so scary for a child when adults are frightened
like that and offer no reassurance."

Through a storm of what seemed to be a torrential rain of
rocks, debris, and water, Gervasia ran to the top of the village,
almost by instinct. "It was dark. There was no power. Water
was falling from everywhere but the worst thing was the roar,"
she recalls. "I dreamt of it for many months afterwards. I would
wake up suddenly and remember that roar—the most terrifying
thing I have experienced all my life."

Gervasia's grandmother and brother survived but her
mother, uncle, and many schoolfriends did not. Her mother had
gone to get the family's livestock off Mount Toc, as instructed
by officials at the dam, and had been staying overnight in the
village of Erto. Her uncle, an ENEL employee on the dam, was
one of fifty-eight staff who lost their lives in the disaster. With
her father already dead, her uncle had been a surrogate parent.
Now Gervasia was an orphan. "Suddenly I knew what it was
like to lose everything in just one night. My life shifted from the
blue Alpine sky of Longarone to a bleak orphanage on a foggy
plain near Venice."

Gervasia was not alone. Some 2043 people, including almost
everyone who was in central Longarone that night, were dead.
The reservoir had been almost completely emptied by the landslide. Above the millions of tons of rock was the exposed, sheer,
almost white rock face from which it had all slid. It formed a
huge letter M shape more than a mile wide, sliced from the face

Today the Vaiont Dam remains a spectacular piece of engineering. The dam resisted forces it was never designed to withstand when a landslide forced virtually its entire reservoir over the crest, drowning 2043 people.

of Mount Toc. Yet incredibly, at the far end of what had been the lake, the dam remained standing. There was minor damage on the crest near the left abutment but nothing else. The dam was believed to have withstood a force of four million tons—eight times that which Carlo Semenza had written into his design specifications. With tragic irony, the worst dam disaster the world had ever seen had been a brilliant structural success.

OVER THE NEXT YEAR THERE WERE VARIOUS INQUIRIES INTO THE CAUSE OF the Vaiont Dam disaster. The Ministry of Public Works appointed Carlo Bozzi to head a commission; ENEL appointed its own team of investigators under Marcello Frattini, and the Italian parliament began hearings. In January 1964, the Bozzi Commission blamed a combination of bureaucratic inefficiency, the withholding of alarming information, and buck-passing among top officials. Italian Prime Minister Aldo Moro suspended a whole raft of officials pending criminal investigations.

In December 1967, more than four years after the disaster, the public prosecutor of Belluno, the district authority, charged eleven men with responsibility for the disaster, the indictments ranging from manslaughter to negligence. After a trial lasting more than a year, in which the court heard from more than twenty-five hundred witnesses, three engineers were found guilty. Two served short jail terms.

But nothing seemed conclusive. The whole national debate, fuelled by the benefit of hindsight, seemed to be as informed by the need to apportion blame as to establish any truth or safety guarantees for the future. Perhaps the tragedy was too great, the warnings too stark for the truth to have a chance. Yet over time, the importance of the disaster and interest in it did not diminish. Where the criminal proceedings halted, academia took over. More than fifty scientific papers on the Vaiont disaster were published over the next decade.

For structural engineers, one conclusion was obvious. A much more complete understanding of the geology of an area that was to be dammed was as important as design and structural calculations. The question was how to get such a complete picture. Now that the slide at Vaiont had happened, now that geologists could see inside the mountainside, could Mount Toc yield clues as to how understanding of an area's geology could be obtained in future? Edoardo Semenza had warned this was a prehistoric landslide site and another slide had, as he claimed, been waiting to happen. Why exactly? Could it have been foreseen?

In 1975, Frank Patton, a consultant engineering geologist, became very interested in Vaiont. He had been asked to evaluate a large landslide in the proposed reservoir of a dam site in Canada. There was an obligation to look at precedents, but the various conclusions in Vaiont were confusing and contradictory. The following year, accompanied by a colleague, Professor Alfred. J. Hendron Jr. from the University of Illinois, he flew to Italy to look at the site personally. They wanted to look at what geologists termed "the failure plane," its geometry as well as its geology, and the water pressures that would have been acting on it. "At that time very few of the people who had written papers had actually been to the slide and fewer yet had gone to the failure plane of the slide and determined what was there," Patton recalls.

For Patton and Hendron certain things just did not add up. Müller had said that the prehistoric landslide area sat on a layer of hard rock and was shifting slowly, somewhat like a glacier. If it was, and was subject to low-level water pressures from the reservoir water, the slide should not, in their opinion, have happened. "These essentially were incompatible facts and certainly did not explain the various periods of instability," Patton claims. "There were three periods of instability: the prehistoric slide, the large movement and perimeter crack in 1960, and finally the major slide in 1963. Any explanation had to account for all three events."

Detective work at Vaiont was now much easier. With the rock from the slide in what had been the reservoir and the "failure plane" from which it had slipped exposed to view, Patton and Hendron, equipped with walking boots, ice picks, and geologists' hammers, climbed into the mile-wide, M-shaped gash in

Mount Toc. They measured the inclination of the rock beds at the back of the slide and compared it with the inclination of the rock beds at the base of the slide. Patton's description: "This was a peculiar slide, twenty-five to forty-five degrees at the back and dipping away from the dam some ten to twenty degrees at the bottom, giving a chair-like shape at the base of the slide."

It quickly became clear there was a layer of clay at the base of the slide. And not any old clay. This was plastic, low-strength; in fact laboratory tests were to show it was the weakest of clays. And it was definitely a layer. Patton and Hendron checked at ten, twenty, thirty, then forty and fifty locations across the base of the mile-wide slide. The thin layer of clay was a constant: sometimes four inches thick, sometimes less than half an inch. This was particularly surprising. Leopold Müller, the pioneer of geomechanics who had made so many studies of the mountain, had published a report after the slide stating that there was no clay here. Yet there obviously was and its presence changed everything.

Frank Patton now manages a company specializing in borehole instrumentation in Vancouver, Canada. Nearly a quarter of a century after his original research trips and thirty-six years after the original disaster, he agreed to travel halfway around the world and return to the Piave valley with a television crew from The Learning Channel. He would show viewers exactly what he meant.

Today the dam still rises magnificently from the sides of the gorge, with its elegant inward curve and narrowing profile as it descends to the foot of the gorge complementing the slopes and peaks that surround it. But there is no water behind it. Today Carlo Semenza's masterpiece holds back nothing except the thousands of tons of rock and shale that displaced all the water on the fateful night Longarone died. On its right-hand bank, the sheer, solid rock-slopes that form the gash left in Mount Toc by the slide rise up to the summit. Light vegetation, including young conifers, has sprung up wherever it can put down roots in what still seems at first glance to be nothing but a mass of light gray rock.

Getting to the slide debris is no cakewalk—it involves scrambling over seemingly endless rocks and boulders. But it is amid this rubble that Frank Patton can demonstrate the clue he found in 1975 on his first trip. Picking up a rock half the size of the palm of his hand, he rubs away a gray substance—clay. But in the failure plane, further up, the clay is still in place. The uphill climb here is even steeper. Rocks still occasionally roll down the mile-wide gash in the mountain as they did on October 9, 1963. But halfway up, Frank Patton has no problem finding what he is looking for, the key to the Vaiont disaster: a gray layer that streaks part of the void left by the slide. "It's soft, soft enough for you to push your finger into it," he says, doing just that. "It's the same fat clay—high plasticity, low strength. When we first found it we continued to excavate. We found it all over the place."

Sitting on a rock nearby, Patton explains how the clay would have acted as an impervious membrane or a dam inside the mountain. But did this matter? Could it help explain the sequence of movements and slides before the disastrous one in 1963? Water pressure from the reservoir had seemed to be the key factor in controlling the rate of slide. Müller and the other geologists had advised steadily lowering the levels if the movement became too pronounced and it had seemed to work—until the final fatal slide. But the reservoir filled fastest during and after heavy rain. Was it the filling of the reservoir or was it the rain that caused the movement? Could rainwater draining inside the slide area, particularly after the wide M-shaped crack had appeared after the first slide in 1960, have affected the stability of the mountain?

Patton and Hendron had plotted the rainfall levels against the known movements of the mountain. "We made several plots: one plotted for a given reservoir level [versus] the rainfall that had occurred in the previous week; a second plotted the reservoir level versus the rainfall that occurred in the previous two weeks, then thirty days previously and finally forty-five days previously," recalls Patton. There was little doubt that over three years there had been most movement when there had been most rain. But why? Was it pressure on the outside of the mountain from the increased levels of water in the reservoir, or was it pressure from inside the slide area where water might be collecting?

Knowing about the clay, Hendron and Patton suspected that rainwater had got into the mountain and drained down to the bottom of the slide. Unable to escape because the clay acted like a dam, water pressure built up inside, both below and behind the clay. The result would be an enormous force below the clays in the base of the slide. But how to prove this? Water-pressure measurements from inside the mountain were the only option. Patton and Hendron checked the records and discovered with excitement that there had been a set of water-pressure measurements taken over a few months in 1961. But it turned out that those records had been lost.

On their second trip to the Vaiont Dam, Patton and Hendron teamed up with Edoardo Semenza and visited the slide area. The Americans had a lasting respect for Semenza. No one living knew more about the geology of the area. No one, as far as Patton and Hendron were concerned, was less biased. Semenza's reports warning of the dangers had been written before the disaster. So many of the reports or papers written afterward by those who had done surveys before the disaster seemed to be colored by self-justifying hindsight.

Indeed, the more they studied Semenza's reports the more credible they seemed. Patton and Hendron believed that what the Italian geologist had identified as mylonite was the very clay they believed was crucial. "Once we had visited the site, we realized that the two terms were essentially identical here. He did describe his mylonite as being a clay-rich mylonite at the base of this slide mass he had identified," notes Patton. "It should have alerted people."

But while comparing notes with Semenza, an even more important clue emerged. In 1960, as part of his survey efforts, Edoardo Semenza had made four boreholes. He was collecting straightforward geological information rather than water-pressure data but, having inserted steel pipes in all four holes, it became possible to measure ground-water levels in each of these. There was one problem. Because of the chair-like shape of the slide plane with its deep straight back, three of these boreholes did not penetrate to the base of the slide. These three boreholes gave water-pressure level readings of thirty-three to ninety-eight feet above that of the outside reservoir levels. That was what Patton and Hendron would have expected in rock adjacent to a reservoir like this.

However, one borehole did penetrate to the base of the slide, cutting through the layer of clay at the bottom. It gave an extraordinary reading. The water pressures here were 230 to 295 feet above the reservoir level outside. "It was an enormous difference...equivalent to the pressure of a twenty- or thirty-story building," recalls Patton. "There was a complete imbalance of water pressure. Very high pressures acting below the clays, much lower water pressures above the clays."

For Patton and Hendron this was the only proof there would ever be that the clay was holding back rainfall and snow-melt inside the mountain. It was certainly something Müller had suspected, hence his original direction that SADE should build two drainage tunnels at the top of the slide. The tunnels had encountered little water, thus water pressure did not seem to be much of a problem. In fact, water pressure was an enormous problem. Drainage, done properly, might have saved Longarone. The problem was the tunnels were in the wrong place, according to Frank Patton. "The tunnels were just too high on the slide—inserted in a location where you would have anticipated no ground water. If the adits had been hundreds of yards lower under the base of the slide they would have encountered the high water pressures which all the evidence indicated was there."

Patton and Hendron believe that the difference in relative water pressures was crucial in the final disaster sequence. In the week leading up to it, the Italian engineers had resorted to what had worked before: lowering the water in the reservoir. This was Leopold Müller's solution and was based on a simple premise, trying to limit any slide. A rapid lowering of the reservoir would reduce the water pressure holding the slide in place, and dislodge what was known as the toe or lowest area of the slide which would fall into the gorge and serve as a buttress to hold up the rest of the mountain. It seemed a risky business but it had apparently worked on several occasions after the first major slide in the autumn of 1960. A rapid lowering of the reservoir had arrested the movement of the mountain. The slides had not been stopped but had been managed.

In September 1963 the reservoir had been at its highest-ever level: over 787 yards. When the movement of the mountain had accelerated, the dam engineers had tried the usual trick— releasing water. Throughout the first week in October they lowered the water level in the reservoir. Yet now, with the rain having poured down for days, the water that had penetrated through the cracks in the mountainside was building up rapidly inside the slide; the pressure was growing. The water level on the other side of the mountain—in the reservoir—was sinking, the pressure declining. "While there was some form of equilibrium of pressure the slide was likely to be held in place," Patton concludes. "Given the water that was in the mountain, rapidly lowering the reservoir was the worst thing they could have done." It is just the last of many ironies that the remedy at the Vaiont Dam may well have triggered the disaster.

As the mass began to slip, a factor no one had predicted came into play. Instead of the slow, controllable, glacial-like slip Müller had predicted the slide mass quickly gathered an alarming speed. Having been the cause of the slide in the first place, the water and clay in the mountainside now compounded the impending disaster by acting as lubricants. Experiments done as recently as 1997 on samples from the slip surface demonstrated that the slide surface of the mountain lost as much as 60 percent of its static sheer strength or frictional resistance when the rate of slip exceeded four inches a minute. The faster it went, the faster it could go. As friction became minimal, the slide was able to move a distance of 437 yards at up to 33 yards a second. The speed at which it all happened was the last piece of the equation that ensured massive disaster.

PERCHED ON A ROCK, LOOKING DOWN ON A MAGNIFICENT BUT REDUNDANT piece of architecture, a dam that dams no water, Frank Patton says the Vaiont disaster was a product of its time and circumstance. "It was a period of transition. There was the ability to design and construct very high dams but the science of rock mechanics, the understanding of these large rock masses, was in its infancy." The only consolation is that the Vaiont Dam disaster served as a major catalyst to learn more. "The information gained here has been used to stabilize a number of slide masses on dam projects around the world. That's the main reason we've made all of the information we've collected available to others."

Edoardo Semenza has never given up his search for the whole truth on the Vaiont Dam disaster. It has defined his career, his life, with numerous field expeditions to the site of the disaster, academic papers, professional conferences and seminars. His house contains a mass of paperwork, plans, and photographs going back to his earliest work, his earliest warnings. For the whole picture, the whole truth that has emerged today is in the end an extension, an expansion of the truth he articulated forty years ago. Today, there is one key question for him: if his seniors had paid attention to him could the catastrophe have been avoided?

"To understand all that, you need to appreciate Italian university mentality. Old professors believe the young don't understand anything," he complains, flicking through the album of black-and-white photographs that show Mount Toc before the slide. "Some colleagues of mine still behave like that. I don't, I value what my young assistants think." Like Frank Patton, he believes his photographs, taken in 1960 as part of the first survey, leave little room for doubt. "I was able to understand the real nature of the area because I surveyed the mass from many sides...on the northern side the layers are almost regular but from the western side you can see all the characteristics of a landslide moving away from its base."

Edoardo Semenza closes the photo album with an air of resigned formality. Half a lifetime on, the man who gave the most complete and persistent warning of what would happen in the Vaiont Gorge still feels the tragedy personally. He sighs. "No one believes the pain I feel." No one who hears him say this can doubt it.

11

WALKING ON WATER

Alexander Kielland

IF EVER A MAN-MADE STRUCTURE CAME CLOSE TO THE MIRACLE OF walking on water, the *Alexander Kielland* would seem to be it. In the offshore oil industry it was known as a "semi-submersible," a ten-thousand-ton steel oil rig with a difference. Its five legs, each nearly twenty-eight feet in diameter, floated on five huge circular pontoons when moving from location to location, but were filled with water and semi-submersed over sixty-nine feet into the sea for maximum stability when at anchor. On its completion in 1976 it seemed a triumph of structural engineering for its makers, Compagnie Française d'Entreprise Métallique (CFEM), whose pentagonal, peripatetic design seemed a breakthrough for the industry. Its stability was its major selling point.

The shifting sands of the offshore oil industry were to determine otherwise. The first blow was its use. Designed to explore for oil throughout the North Sea with permanent platforms erected where it struck black gold, the *Alexander Kielland* found its role downgraded from the start. It became a "floatel" for its Norwegian owners, housing the workforce of more advanced rigs who were ferried in both day and night by helicopter.

A mass of common rooms, dining areas, and two cinemas surrounded a maze of modular sleeping quarters all grafted onto its steel platform. By March 1980 the *Kielland* was playing host to more than two hundred workers in the busy Edda oil field, the last of seven oil and gas fields in the Ekofisk development nearly two hundred miles off the Norwegian coast. For the workers, the *Kielland* was a home away from home, a refuge from the noise and dangers of the rigs, a place to relax, socialize, and forge the deep friendships common among men working in such perilous industries a long way from home.

As March 27, 1980, dawned, it promised a stormy day. It was nothing exceptional for the North Sea, but the swell and wind were enough to play havoc with the helicopter shuttle service. Unable to get off to their destination, Gunvald Falk, a Norwegian contractor, and his friends killed time playing cards, drinking coffee, reading. Others, like Ted Brooking, an electrical consultant from Scotland, were unable to get back to their usual floatel and had to be dropped off on the *Kielland.*

By 6 P.M., a wind of fifty-two to sixty-five feet per second and waves of up to thirty-three feet were shifting the *Alexander Kielland* away from Edda 2/7C, the drilling rig to which she was connected by means of a gangway. The link was hauled in. Ted Brooking, having had dinner, settled down to watch Robert Redford in *Jeremiah Johnson.* Gunvald Falk was in the reading room perusing the latest newspapers.

"It was getting dark and we heard a sort of crack and everything shook a bit," recalls Gunvald. "I said to myself, 'Gosh, the waves are high tonight.'" Ted Brooking, two or three minutes into his film, heard the same thing. "There was a loud crack, that's the only way I can describe it, quite a loud crack, a bit of vibration." The cinema audience was quite startled but seemed to decide collectively that there was probably nothing wrong. Until about a minute later. "There was one loud, loud crack, a lot of vibration, and then the thing just tilted, everyone just tilted to one side. We knew there was something wrong this time," recalls Ted.

Anything that was not fixed to the floor started moving. The cinema was a temporary one with lots of loose piping and other building material stored behind the screen. "These pipes—fairly heavy pipes—started sliding down the floor and carried on sliding," says Ted Brooking. "I jumped back against the wall and let them slide right past me." He realized he had to get out and get out fast. But how? Reckoning he had no chance through the main entrance, he headed for one of the doors at the side of the screen. He followed the wall in what seemed to be complete darkness.

Once in the corridor, Ted saw a square of twilight to his left. It was a hatch leading from the workshop. Using a rubber hose that had served as an airpipe to haul himself up the corridor and through the workshop, he managed to climb up several fixed pipes to get through the hatch. A number of other men were already hanging onto the handrails on the main deck. The *Alexander Kielland*'s list was now so marked that they all seemed to be face to face with the sea swell. But there was no panic or shouting, even when the cause of the trouble became clear. "I could see the leg floating in the water by the side of the rig," Ted Brooking recalls. "My immediate thought was, okay, the leg's broken off but this thing's still up, we'll stop here, and someone will come and get us. You don't think it's gonna sink or capsize—you think it's just gonna stop there."

Gunvald Falk watched tables, chairs, everything fall over and slip toward the wall. An alarm rang; then there was silence. "The lights went out and it was as if everything was completely dead. Then I was thrown against the wall," he recalls. With the wall gradually becoming the floor as the list increased, Gunvald managed to stagger to the door and force it open. "I sort of rolled out into the hallway and everyone rolled out after me, onto me. We made our way toward a light and found ourselves on the open deck. It was only then that I realized the seriousness of the situation. We were so low in the water the sea was washing up onto the main deck."

Although the Edda 2/7C oil platform was within shouting distance, the tilt—now forty-five degrees—was so severe that capsize seemed inevitable. Certainly the radio operator on the *Alexander Kielland* thought so. At 18:29 he sent out an emergency message on Channel 9: "Mayday, Mayday. *Kielland* is capsizing." For those on or near the open deck, the rig's seven life rafts were the only obvious hope. Ted Brooking began to make his way to them, passing those clutching the handrail on the platform. As he passed hand over hand, the angle of his climb got ever steeper. It was clear by now that sea water was rushing into the rig as vents, windows, and doors came level with the sea.

There was one problem: the life rafts were never intended to be launched from a rig listing at forty-five degrees. The hooks on the lowering cables would simply not release at this angle. Even if they did, it seemed likely the boats would hit the twenty- to twenty-six-foot waves at the same angle as the list of the platform, capsizing or being crushed against it. Gunvald Falk, having managed to grab the last life jacket from a chest on the deck, passed one of the life rafts as he scrambled to get to the *Kielland*'s highest point, the B column.

"I heard somebody shouting: 'You have to come now, Gunvald! We are launching!' I took one look at the angle of the boat and just thought, no chance." He was right. Although this life raft was one of the three that did manage to launch, it was quickly smashed against the side of the rig and overturned.

Ted Brooking, on reaching a life raft ready for launching, came to the same conclusion as Gunvald. "It was only fiberglass....I took one look and thought, well, if you think that thing can get launched into that...I didn't fancy it at all."

The only alternative was the icy water. Although survival times would not be long, it seemed they would not need to be. Ted could already see the crane operator on the Edda rig using the basket on the end of his crane to pluck survivors out of the water. "I didn't have to jump or anything like that. She just started to go, slipped under and we went into the water with her," Ted remembers.

As the *Alexander Kielland* lost all stability and slipped under, Gunvald Falk decided he had no choice and dived in. "The life jacket dragged me up and when I reached the surface I saw what had happened. There was the rig, the four legs or globes floating upside-down, overturned, capsized."

Clinging onto some wreckage he had found floating in front of him, Gunvald managed to make his way toward an enclosed overturned lifeboat. With the help of several other men he managed to right it, open the hatch, and scramble in. Neither the motor nor the radio would work so they just drifted through the fog and the swell waiting to be rescued. Every now and then they would hear shouting or knocking on the side of the boat and would open the hatch to try and haul another half-dead survivor into the relative comfort of their lifeboat cocoon.

Ted Brooking found himself sucked under several times after hitting the water. "The second time I thought, well, I'm in trouble now. I thought I was actually going to die at that moment." When he managed to break the surface for the second time he looked for something to hold on to and saw a black PVC cover containing an inflatable life raft. By synchronizing his swimming efforts with the waves, Ted managed to reach the life raft and, with two other men who were struggling in the water, opened it.

Inflating it was another problem. It seemed to inflate only partially on opening. Unable to get in, Ted helped push one of the other men, a Norwegian, aboard. "He's kicking and I'm shoving him and he just flopped in. Then I heard a rush of air...air escaping, right? No. It was inflating. The thing just opened up while I was still holding on to the side." The Norwegian managed to help Ted aboard. It was just in time. He could no longer feel his hands. It was only a matter of moments before he completely lost his grip on the painter rope around the life raft.

Ted Brooking was lucky. Within twenty minutes the large orange canopy that formed the tent-like cover of the life raft had been spotted by an oil-supply ship. But the rescue would be almost as perilous as the slide into the water. "We could see the propellers of the ship turning, hear them turning as the ship approached. I started to think one freak wave and we'll be sucked into the blades. I was worried, panicking as well because it was out of my control."

They managed to get the crew to approach the life raft from the other side, then they had to make the terrifying leap from its relative safety into the rescue nets slung over the side of the ship. But they made it and, safely aboard, Ted Brooking and his companion were placed in bunks and wrapped in blankets as the ship sailed on to look for more survivors. "I just lay there in the bunk and I'll tell you, I've never been so cold in my life. I felt myself thawing out from the top down...all the way down."

Gunvald Falk was party to a similarly dramatic rescue. Six or seven of the twenty-seven people on board his life raft managed to scramble aboard a supply ship that pulled alongside shortly before midnight. However, rammed together in the stormy seas as it tried to pick up the others, the supply ship's hull cracked the roof of the life raft. The ship and the life raft parted and lost each other in the fog. It was 3:30 A.M., eight hours after the *Alexander Kielland*'s collapse, before a helicopter spotted Gunvald's crippled craft.

With the supply ship standing by, its spotlights blazing, Gunvald Falk had to jump into the water to get hold of the rescue basket as it swung down to the life raft. However, he never actually managed to get in, finally being winched to safety clinging to the base. "They told me afterward that they had wanted to release it because they could not see me. I was three or four meters out of the water before they spotted me trying to climb over the rim of the basket."

As the rescue helicopters landed and neighboring ships and rigs reported the numbers of survivors picked up, the scale of the disaster became clear. Just 89 men on the *Alexander Kielland* that night had survived. Some 123 had died, either trapped in the rig or frozen in water where the survival time was less than half an hour. It was by far the worst-ever accident in the offshore oil industry's history and in the cold spring daylight the cause of the disaster seemed clear. It was exactly what Ted Brooking had seen in the dark before he slipped off the rig into the water.

The underside of the huge circular pontoons that had supported the legs of the *Alexander Kielland* were clearly visible—the rig was upside-down. However, there were only four of them. The whole of the fifth leg, pontoon and limb extension, lay nearby. Detached and horizontal, it bobbed up and down uselessly in the swell. For some reason there had been a catastrophic structural failure on the rig. With ten other pentagon rigs in use, the offshore oil industry would need quick answers to the only real question: Why? Was it an explosion? Metal fatigue? A collision? Even sabotage? And whatever it was, were the other rigs safe?

THE NORWEGIAN GOVERNMENT QUICKLY APPOINTED A FIVE-MAN COMMISsion to come up with some answers. Headed by a district judge, Thor Naesheim, it included one of the world's foremost experts on the design of offshore platforms, Professor Torgeir Moan. The first job was obvious: salvage the evidence and interview the witnesses. The drifting detached leg, column D on the platform's design plans (see Figure 10), was towed to the Norwegian port of Stavanger within three days. The inverted rig took a little longer, eventually being harbored in a fjord north of Stavanger when it arrived three weeks later.

All the witnesses, survivors from the *Kielland* and those on the Edda 2/7C platform, were quickly interviewed. There were no substantial discrepancies. The loud cracking noises most heard were of course perfectly compatible with the separation of column D in the storm. A collision seemed impossible. No one had seen or registered a vessel nearby, and the *Kielland* had been moving away from the Edda 2/7C fixed platform at the time, the main reason the gangway had been hauled in.

Within four days, Professor Moan and his colleagues were clambering over the severed leg in Stavanger harbor, where by sheer good fortune another pentagonal semi-submersible was docked. Making useful comparisons, investigators soon noted that the key feature of the severed leg seemed to be a

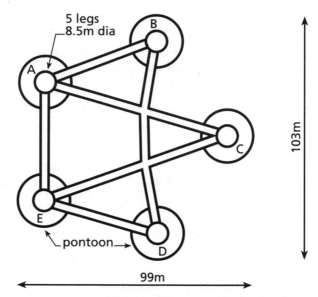

FIGURE 10
The pentagon design of the *Alexander Kielland.*

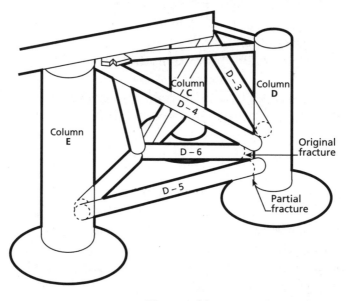

FIGURE 11
Close-up of columns C, D, and E, showing bracings.

sheared section of a horizontal steel brace, D-6, which was
still protruding from the leg. It was one of four lower braces
which, along with two upper braces, connected the leg to the
rest of the platform (see Figure 11). All these braces were essentially hollow steel pipes. Their key feature was their size:
the lower braces were eight and a half feet in diameter. In
other words this was a pipe in which the tallest of men could
walk comfortably.

It was on the remains of this brace that the official investigators led by Torgeir Moan now focused their attention. A fatigue crack—a crack that forms and grows in a gradual and
progressive fashion—had evidently spread in two directions
around the whole circumference of the brace. The crack may
have been growing for years, each wave—or loading, as structural engineers would say—extending it by a few millionths of
an inch. But with millions of wave motions a year that would
have been enough. "There were very clear marks on the surface of the metal, like the growth rings in trees, left by the repeated actions of the waves," notes Moan. "There was no
doubt about it: this had been a fatigue failure. The metal had a
smooth surface. With a sudden break you get roughness."

The next question was what had caused the original crack.
Analysis showed that the fatigue life of the broken brace was
about seven years; that of the other braces was a little more.
However, all had a lifespan of less than ten years. But although
that in itself was alarming, it was not the key. Where the crack
had formed there were the remains of a weld. After the construction of the rig, a small cylinder used to house a hydrophone—an underwater microphone used to pick up signals
from seabed transponders to help position the vessel—had
been attached to the underside of the D-6 brace. Right from the
start, the welds around the hole cut to insert the steel tube on
which the hydrophone was mounted were poor, investigators
concluded. A crack of at least three inches had appeared during or shortly after the welding itself.

How could they be so sure? Leading experts in the field of
forensic engineering investigation, Failure Analysis Associates

(FaAA) in California, had been brought in to advise. Their chief metallurgist, Charles Rau, was convinced by three crucial clues.

Firstly, there was the nature of the crack. "This was what we call a cold crack—that is, a crack formed during the cooling down from the hot welding process. It has distinctive features, so from its basic shape and the appearance of the fractured surface we know it was formed as part of the solidification process."

Secondly, looking into the fracture, Rau and his fellow investigators decided that it had begun to corrode before the *Alexander Kielland* went into service, before its anti-corrosive system—Cathartic Protection—had been applied. It was a crack that had its origins in the yard.

Thirdly, and most tellingly, there was the rig's paint. What convinced the investigators was that the original crack had several layers of paint inside it, along the lateral inside edge of the metal. These layers were in the same sequence as those on the rest of the rig and sported paint of the same chemical composition. In other words, by the time CFEM had come to spray-paint the rig—first red, then black, then green some time after the welding had been done—the crack was in place.

What engineers term continuity or lack of continuity was the key in the defective weld. "When you weld two pieces of metal together, you assume you are getting a continuous seal. Welding is no different from gluing," notes Piotr Moncarz, principal engineer at FaAA. "However, if you have a defect in that glue, a discontinuity in that glue, it will act like a notch from which a failure, a tear, could start." The analogy Moncarz uses is a simple piece of paper, with and without a tear.

"If I take a piece of paper and pull on it," he says, tugging on a sheet of foolscap paper from both sides, "that's a pretty strong piece of paper. Now take the same sheet of paper and make a small tear in it. If you pull with the same force on that sheet of paper, it will tear—starting from that original tear. This is the same phenomenon that occurs when you have a structure with a defect. The defect becomes the crack, the crack becomes the fracture."

But why with one brace out of six fractured would the leg detach from the *Alexander Kielland*? And even if it did, why could the rig not have remained afloat with four legs, or at least afloat long enough for everyone to get off? After all, four was the number of legs more conventionally designed rigs used. The whole point of the pentagon was the improved stability afforded by its five legs and, indeed, the improved structural redundancy in the event of failure. That is, if one leg became non-operational—if, for instance, water flooded a pontoon—the other four legs were designed to provide sufficient support. What had gone wrong with the *Alexander Kielland*? And could it happen to any of the other pentagons?

Model tests and structural calculations based on the *Kielland*'s design specifications soon established that once the D-6 bracing had broken off, the five other braces securing the leg to the platform quickly followed. They were simply overloaded—the structure had no redundancy built into it to allow for such an eventuality. Indeed, without D-6, the strength of the remaining five bracings supporting column D was barely enough to withstand what engineers termed "still-water loading"—the load applied in calm water (see Figure 11).

The two upper braces that secured the top of the *Kielland*'s column D to the deck were torn from their sockets. The other lower bracings had severed at various distances from the actual leg. "Damage tolerance" was negligible in the pentagonal design simply because—incredibly—there were no mandatory requirements for such tolerance in offshore oil industry structures at the time.

As for the bracings, so for the rig's columns or legs. Without the bracings, the loss of column D, one of the pentagon's five legs, was simply a matter of time. "The initial stability calculations did not even consider the possibility that the rig might lose one of its five legs," the Norwegian commission concluded. "Neither did any regulation require stability for such circumstances at the time when the *Alexander Kielland* was approved." Ironically, although there was plenty of built-in structural redundancy for the non-operation of a leg—ballasting problems, for instance—there was none that ever allowed for the severing of a leg. It was simply an unforeseen circumstance.

And once a leg was lost, a severe list to about thirty to thirty-five degrees was inevitable, the commission decided. At this angle the rig stabilized somewhat but the capsize sequence was already in train. With no watertight seals, the sea flooded into the lift shafts in columns C and E, then the doors, ventilators, manholes, and windows. "The deck heels, water gets in. Heels more. More water gets in," explains Piotr Moncarz. "Eventually it reaches the point where the buoyancy system of the platform can no longer stabilize the rocking rig. You reach thirty-two degrees and the platform goes flop. Suddenly it's flipped over and is upside-down—it's achieved its new stable position, legs in the air." The whole sequence took all of twenty minutes.

The official inquiry into the *Alexander Kielland*'s capsize found a number of factors responsible for the disaster. It blamed CFEM for the design and construction failures, notably poor stability, lack of redundancy, and the faulty weld itself. It blamed the Norwegian classification society Det Norske Veritas that registered the vessel, and the Norwegian Maritime Directorate for allowing the rig to leave the yard with a potentially fatal flaw as well as for not identifying the crack in regular inspections afterward. It blamed Phillips Petroleum, the rig's operators, for the poor standard of safety and emergency training of those on board, the failure of safety equipment, and the tardy arrival of rescue craft.

But the beginning of it all had been the "unfortunate design, dimensioning, and material quality of the hydrophone support as well as its connection to the bracing." The reason for this, the Commission of Inquiry concluded, was that the hydrophone support was "considered as outfit and not as part of the load-carrying construction at the planning stage." In other words, the weld crack may have been noted by inspectors at the yard, in dock or at sea when divers check offshore rigs. They just may not have considered it important.

"It's my belief that the people who inspected this particular weld did not appreciate that it was intended to be structural.

They thought the hydrophone was something which was just being stuck on with chewing gum or putty—an attachment," says metallurgist Charles Rau. "I suspect, although we'll never know for sure, somebody in the chain of authority decided not to be as rigorous in enforcing the codes for structural welds because they judged, incorrectly, that this was not a structural weld and that if it fell off, it would be of no consequence with regard to the structure."

Despite its convincing argument, the official report from the Norwegian commission simply served to fuel rather than close the debate about what had happened to the *Alexander Kielland*. Although no one denied the central thesis of fatigue failure, many came to believe that other factors, ignored or skirted over by the commission, had played a significant part. There are fatigue cracks in most rigs, critics complained, but most do not topple over in less than thirty minutes as the *Kielland* had done.

"They just jumped to a conclusion on seeing this fatigue failure, and all within a few days of inspecting the rig," says Kristian Reme who, after losing his brother in the disaster, came to head the Kielland Foundation, an action group for relatives. "That's where I still believe the commission got it wrong. I'm not saying that their theory is wrong, but I am saying it's still just a theory because they didn't investigate the other options."

Reme's argument was given some credence by further research. The British government, with three pentagon-type rigs in its oil fields in the North Sea at the time of the disaster, had a particular interest in their safety. London carried out detailed shallow-draft inspections and conducted a number of tests on stability. The French did likewise. Two years after the disaster, with the Norwegian commission's report now published, the French designers and builders of the *Alexander Kielland* asked a French court to appoint a group of independent experts to examine the cause of the disaster. Both inquiries produced some unexpected results.

Computer-modeling research carried out by the British government and Lloyd's Register concluded that the pentagon design was particularly prone to fatigue in one line of bracing when subjected to waves of a particular size from a particular direction. The vulnerable line of bracing was the cranked member joining columns D and B. It traversed the full width of the rig and was the only bracing that traversed two other bracings rather than one.

The models determined that this bracing suffered what structural engineers term severe longitudinal stress when wave direction was along the line from column B to D and wave length (the gap between the waves) was about double the 266-foot spacing between the two legs. Analysis of weather reports and logs suggested that these were exactly the conditions on the evening of the disaster.

This effect in pentagon rigs had long been recognized by the offshore oil industry. As a result, such rigs were normally anchored with column C headed into the prevailing wave direction—usually northwest in most of the North Sea. However, because of the way it had to be positioned in relation to the Edda 2/7C platform, its host, the *Alexander Kielland* was apparently moored with column B headed into the prevailing waves.

Britain's Department of Energy and Lloyd's Register concluded that such an arrangement becomes critical if combined with the sort of fracture-fatigue defect highlighted in the Norwegian commission's report. One solution was additional bracing, closing the gaps in the horizontal links by bracing column D to C and C to B (see Figure 10), exactly the remedial action performed by CFEM on one of the *Alexander Kielland*'s sister pentagons, the *Henrik Ibsen,* shortly after the disaster.

At the core of the British research was one simple fact: if the rig had been used for its original purpose, drilling and exploring, rather than for accommodation, the disaster might not have happened simply because it would not have had to be moored as it was. The British had come to that conclusion as a result of the positioning evidence. In fact, the change of use element was something the French builders and designers had pointed out all along but from a different perspective—superstructure overload and its effect on stability. Immediately after the disaster CFEM claimed that the rig's 131-foot high, two-hundred-ton derrick and all its extra accommodation units—off-center above the leg that gave way—would have had a serious effect on stability once the failure sequence was under way.

It was something highlighted in the French experts' exhaustive report which was finally handed over to the Tribunal de Commerce in Paris in July 1985. The French experts pointed out that by the time of the disaster the owners were operating a different structure for a different use from that for which it had been designed, yet had made no changes to the operating manual. They claimed the rig could have remained upright on four legs by the simple act of ballasting during the twenty minutes it was listing at thirty to thirty-five degrees. Pushing a few buttons in the control room was all that was necessary. They concluded that given the operational incompetence, the fatal crack was probably provoked during "exploitation" of the rig rather than during construction.

Overloading of rigs was something both the Norwegian and British governments had lighted on in the wake of the disaster. All seagoing vessels have a habit of putting on weight over the course of their lifetime—additional machinery, equipment, stores, even new layers of paint. All the research showed that while overloading that could result from this "creeping growth" in payloads has little effect on the stability of an intact rig, it could be critical in a rig that has been damaged.

On the face of it the *Alexander Kielland*'s actual declared payload, some 1033 tons, was only half its capacity of some 2100 tons. But that assumed the declared payload at the time of the accident was correct. It also assumed that what was known as eccentric loading—the distribution and height of the four-story-high accommodation modules and the wind forces they attracted—had not made the rig inherently dangerous.

Surveys of rigs throughout the North Sea showed that "unsuspected overload," as it was termed, was a real problem. On one Norwegian rig something like two-thirds of the supposed fifteen-hundred-ton load-carrying capacity had been taken up by equipment and gear excluded from the normal payload inventory. During the early 1980s the British carried out a series of inclination tests on some thirty rigs of every type of design in their sector of the North Sea. The results were alarming. An annual check on weight became a mandatory part of inspections in all sectors of the North Sea soon afterward.

Meanwhile, the suspicion grew, especially among some of the relatives of the victims, that contraventions of loading and operational regulations on the *Alexander Kielland* were so serious that the rig might have been considered unseaworthy at the time it capsized. Ironically, the final answers on loading, stability, and state of repair could have been provided by the *Kielland* itself. Who needed to do computer simulations or laboratory experiments with models when you had the real thing? This was a unique collapse. The structure itself had survived.

After the rig was towed to a fjord thirty miles north of Stavanger, debate raged as to what to do with it. Public pressure provoked two unsuccessful attempts to right the rig; each time a combination of controlled deballasting, inflatable buoyancy, and winch barges failed to get the structure to the critical angle. A detailed two-month underwater investigation turned up more fascinating facts: the welding of all the hydrophone holders was poor; leg C seemed to be more than four inches shorter than the three others still attached. There was so much to be investigated, so many more tantalizing leads. Some suspected that many interested parties did not want the rig righted. After all, if the rig was unseaworthy, the insurance claim would be invalid.

As time went on, evidence became less important than the thirty-six bodies believed to be trapped aboard the rig. Finally, under pressure from the families of the victims, the Norwegian government agreed to pay the cost—thirty-four million U.S. dollars—of one final righting attempt. During nine days in September 1983, some three and half years after the accident, one of the most elaborate efforts in salvage history was successful. Barges, winch chains, and steel buoyancy tanks got the structure to the 178 degrees at which it would right. Contractor Stolt-Nielsen calculated that it had expended some thirty-thousand

One of several ultimately successful efforts to right the overturned oil rig *Alexander Kielland* in a Norwegian fjord. Two of the rig's remaining four legs are visible.

workdays, including forty-five hundred hours under water on three thousand separate dives, to get it right and righted.

As the search teams picked their way through a tangle of twisted walkways, over a deck coated with seaweed and shellfish, the silt and battered wreckage of the living quarters yielded only six bodies. To everyone's surprise, the cinema from which Ted Brooking had escaped that fateful night in March 1980 contained none. Every item removed from the rig was logged and photographed by the police. They included the rig's log which was freeze-dried in a vain effort to glean further clues about load and ballasting on the day of the disaster. The five members of the Norwegian commission visited the rig and quickly decided that most of their report's assumptions about open doors, hatches, and ventilators had been correct. In their view, there was little else to confirm.

The official investigators were followed by representatives from the builders, CFEM, the insurers, and the relatives of the

victims. The only way to thoroughly examine the legs and bracing was to dry-dock the remains of the rig. But no one wanted
to pay the additional costs—perhaps because no one wanted
to know any more than they already knew. Many who had followed the investigation closely were amazed. "In more than
twenty places in the official report the Norwegian commission
noted the need to right the rig to be clear on certain points,"
says Kristian Reme. "But when they had the chance to do it,
they didn't." Many thought a great opportunity was being
missed. "*Kielland* is a unique fatigue object. It was on station
continuously for its whole three-and-a-half-year life up to the
moment of the disaster," notes one engineer. "It should have
been subjected to a detailed fatigue analysis."

As it was, neither the fatigue tests that would have been of
general use nor the stability tests that might have provided
some specific answers were ever attempted. Instead, in November 1983 the *Alexander Kielland* was towed out to sea,
blown up, and sunk in over twenty-three hundred feet of water.
It was perhaps a fitting end to man's presumption that the pentagonal semi-submersible—safe and stable—was the equivalent of walking on water. Man had a long way to go.

INDEX

ABOUT THE AUTHOR

PHILIP WEARNE IS AN AUTHOR AND TELEVISION PRODUCER WHO DI-
vides his time between London and Washington. His
books include *Tainting Evidence: Behind the Scandals in the
FBI Crime Lab,* a major critique of the FBI's forensic science,
and *Return of the Indian: Conquest and Revival in the Americas,*
an analysis of the revival of Native American consciousness.
His television productions include *The Lockerbie Trail* (Channel 4), an examination of the evidence against the two Libyans
now on trial in the Netherlands; *Mayday* (Channel 4/The Learning Channel), a documentary series examining the causes of
some of the world's worst shipping disasters; *The Diamond
Empire* (BBC), a critical history of De Beers; and *More Than a
Game* (BBC/A&E), an investigative documentary series looking
at issues behind the headlines in sports.